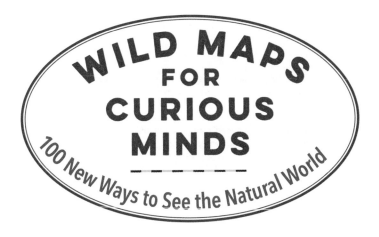

WILD MAPS
FOR
CURIOUS
MINDS

100 New Ways to See the Natural World

WILD MAPS FOR CURIOUS MINDS: *100 New Ways to See the Natural World*
Maps copyright © 2022 by Granta Books
Foreword copyright © 2022 by Chris Packham

Originally published in Great Britain by Granta Books as *Wild Maps: A Nature Atlas for Curious Minds*.
First published in North America in revised form by The Experiment, LLC.

The Experiment, LLC
220 East 23rd Street, Suite 600
New York, NY 10010-4658
theexperimentpublishing.com

THE EXPERIMENT and its colophon are registered trademarks of The Experiment, LLC. Many of the designations used by manufacturers and sellers to distinguish their products are claimed as trademarks. Where those designations appear in this book and The Experiment was aware of a trademark claim, the designations have been capitalized.

The Experiment's books are available at special discounts when purchased in bulk for premiums and sales promotions as well as for fundraising or educational use. For details, contact us at info@theexperimentpublishing.com.

Library of Congress Cataloging-in-Publication Data available upon request

ISBN 978-1-61519-892-4
Ebook ISBN 978-1-61519-893-1

Cover and text design, and maps on pages 44, 48, 71, 74–75, 115, 122–23, 128, 130–31, 134, 142, and 162–63, by Jack Dunnington
Text by Mike Higgins
Illustrations by Manuel Bortoletti

Manufactured in Turkey

First printing November 2022
10 9 8 7 6 5 4 3 2 1

WILD MAPS

FOR

CURIOUS MINDS

100 New Ways to See the Natural World

MIKE HIGGINS
Illustrated by **MANUEL BORTOLETTI**

Foreword by Chris Packham
Ian Wright, consultant

THE EXPERIMENT
NEW YORK

CONTENTS

ANCIENT HISTORY

OUT AND ABOUT

THE WATERY WORLD

GEOGRAPHY

USING AND ABUSING NATURE

EXTREME EARTH

THE PLANET
IN PERIL

THE FINAL
FRONTIER

FOREWORD
by Chris Packham

remember the joy of being able to "sit in the front" of our Ford Anglia, I remember the red leatherette cover and the weight of the book on my lap and running my fingers around the gold-embossed letters that read "Road Atlas of Great Britain." I was "The Navigator," like "Henry the Navigator," like Magellan, Drake, or Frobisher, like a navigator in a Lancaster bomber. I had the very serious responsibility of getting the Packham family from A to B and back using the best, most efficient route. I studied the key, the scale, and the complex matrix of features and memorized all the symbols because I was not allowed to fail. Our lives, or at least the peace and cordiality in our lives, depended upon me and my maps.

You see, my dad didn't get lost. That was unthinkable, impossible, so I couldn't let him get lost on the tarmac ribbons of England, Scotland, or Wales. But there were two things I didn't yet realize in 1967: One, my dad knew exactly where he was going irrespective of my critical deliberations and directions, because he had all the maps in his head already; and two, I would soon manifest the uncanny ability to go somewhere once, anywhere in the world, and then go straight back there without deviation or hesitation, whether I was sitting in the front and paying attention, or walking or cycling, basically connecting with my surroundings. That is surely down to the way my Asperger's mind works with visual information, but probably also down to this formative training in transcribing it into a geo/topographical context using the template of drawn maps. Because I loved maps, I loved memorizing maps, imagining maps, drawing my own maps of imaginary places. Most of all, I loved the Atlas.

The Atlas was the world on paper, the whole wide world spread over pages, spread out on my tiny bedroom floor. I was "here," the jungles of the Congo were there, the altiplano was over there, the desolate outback was on pages 120 and 121. That's where the gorillas were, the vicuñas were, and the Tasmanian tiger might still be. I looked at the map and saw the spaces and places where all the denizens of my animal encyclopedias roamed, and that gave me something I needed: context.

We now live in an age of information. Many people have forsaken knowledge because they can instantly access information. They don't memorize maps because they have satellite navigation, they don't remember facts because they can find them in seconds if they need to. But without knowing things, there is no ability to build context, the structure that links facts together and explains how they coexist, how they relate and lead to one another, because nothing exists in isolation—even on maps of the outback there is no road to nowhere. Like facts, all roads lead somewhere. I would say they go via knowledge to wisdom.

So as much as I will use satellite navigation—it's certainly safer than driving with an atlas balanced on your knees—I don't want to permanently rely on such technology because it will ultimately shrink my world. I like technology to expand the world I know, and when it comes to the study of the environment and the natural world, contemporary technology means we are learning a lot more and a lot more quickly than ever before. And that is irrepressibly exciting. But, and it's a sad "but," we are discovering all these things when so many of the things we are discovering things about are disappearing. So our new knowledge needs to be rapidly, efficiently, and clearly transcribed into a communicable form, so it can be known and shared more widely. In short, this information needs to be made beautiful, and one way of achieving that is through the universal language of maps.

The maps in this book illustrate a very broad range of facts, but they fall into three or four main categories. Many are interesting: The "How deep is the Earth?" map (page 53), where a radial slice from crust to core, transposed onto Canada, the US, and Mexico, puts our planet's size and volume into clear perspective; and where the "tiniest of their kind" of animals live (page 118–19), including the Jaragua lizard, a minuscule 0.6 inch (1.6 cm) reptile from the Dominican Republic, and "Empire of

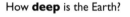

How **deep** is the Earth?

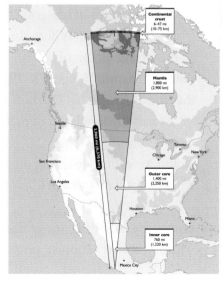

the woolly mammoth" (page 2), which shows the former distribution of the extinct hairy elephant, certainly interested me. The map showing the distribution of the 70 "Moon trees" (page 153) planted from seeds taken to the moon and back in 1971 is amusing, as are the map of countries that have produced the most breeds of dog and the one informing you where you are most likely to

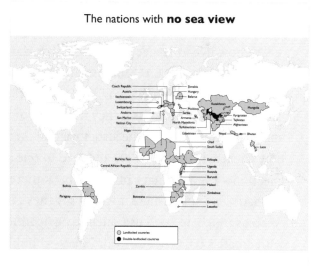

The nations with **no sea view**

be killed by a venomous bite or sting. It seems it might be best to avoid India and Somalia. Some maps are pure trivia, but there are very many maps here that are seriously important because they are shocking or scary. The effects of climate change feature prominently—temperature changes, sea-level rise, winters getting shorter, summers getting longer; as do the impacts we are having on biodiversity. "Who has killed whales since the 1985 ban?" (pages 100–1) is horrifying, as is "Just a few of the hundreds of species that have vanished in the 21st century" (pages 136–37). Some of us will be equally concerned by "Who has kept dolphins in captivity?" (pages 110–11), "Who eats the most meat?" (pages 94–95), and "Where rhinos have been poached" (pages 103). But it's not all doom and gloom of course, and "The hotspots in Earth's crust that could power the world" (pages 78–79) and "The windy stretches that could help power the world" (pages 76–77) show some real solutions to our problems.

If I had the tough job of picking a favorite it would be the "The nations with no sea view" (pages 58–59), because it's the sort of thing my dad and I would have loved learning on those journeys in the long-gone Ford Anglia. Knowledge, you just can't beat it.

CHRIS PACKHAM is one of the UK's top television naturalists and an award-winning conservationist, photographer, and author.

INTRODUCTION

e often hear that we live in a globalized world. Perhaps that's why we increasingly turn to maps of the world to better understand the planet and its occupants. Ian Wright's *Brilliant Maps for Curious Minds*, as well as Matthew Bucklan and Victor Cizek's *North American Maps for Curious Minds*—the predecessors to Wild Maps in this series—delightfully showed that there was little that couldn't be presented in map form. You might not have thought you needed to know which country has produced the most Miss World winners or where the planet's heavy metal bands are concentrated, but it's pleasing that someone else did and created those maps (India and Finland, respectively, by the way—and there are 198 other equally intriguing maps in the Maps for Curious Minds series, so do take a look if you haven't already). Both previous books in this series had chapters called "Nature" (my favorite nature map: "Countries with no rivers"). Those chapters offered the obvious starting point for *Wild Maps for Curious Minds*, showing us new ways to see the natural world—and how the human race is part of it, for better or worse.

So began the happy search for the 100 maps of this book, strikingly illustrated by Manuel Bortoletti. Some we created ourselves, but many existed already, made by academics, bloggers, professional mapmakers, researchers, campaigners, and institutions. The process of contacting the original creators of these maps was one of the joys of making this book. A community of generous and talented mapmakers out there was willing to grant Manuel and me the opportunity to re-present their work here. John Nelson ("Imagine all the oceans as a single body of water"; page 38) was fascinated by the mind-bending map projection of oceanographer Athelstan Spilhaus and the way it reveals what you might call our terra-centric gaze. Climate scientist Brian Brettschneider's blog is a veritable library of fascinating meteorological cartography, but particularly eye-catching were his four maps showing how the seasons of North America have transformed over the last sixty years (pages 146–47). For another weather-related gem, see "It's raining, but is it pouring?" (pages 46–47), a clever spin on precipitation data by data visualization specialist Erin Davis. "All the rivers in the world" (pages 60–61) and "All the lakes in the world" (pages 66–67) are deceptively simple, but built on the expertise of the HydroSHEDS and HydroLAKES projects.

The increasing availability of powerful and affordable digital mapping technology has led to a boom in interactive online tools that helped us create at least two maps: Sam Learner's wonderful website River Runner is behind "The tiny creek that connects the Atlantic and Pacific" (page 39), and we would have found "Australia boldly goes" (page 161) a bit trickier without Chris Yang's interactive creation, "Countries Mapped onto Solar System Bodies."

There are also nuggets of data out there begging to be rendered in cartographic form. Some of the facts we found and turned into maps: "All the private gardens in the UK" (page 29) and "The state of land use" (page 71) might prompt you to look again at the everyday environment that is perhaps taken for granted. Others—such as "How much forest have we destroyed?" (page 142)—are frankly an indictment of our stewardship of the planet.

There is also a lot of fun to be had in moving beyond traditional maps and into infographics that play with scale. As a young child, I was shown a map of southwest England, where we were moving to at the time—surely, I thought, my brother and I would be able to hop from one side of the peninsula to the other in minutes? Alas, no. But that same sense of wonder at the scale of our natural world (and worlds beyond) is, I hope, to be found in "Jupiter's Great Red Spot could swallow the Earth" (page 157), "The

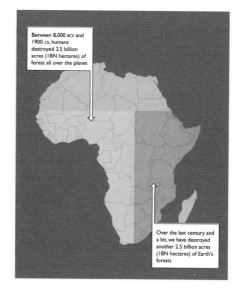

How much **forest** have we destroyed?

Between 8,000 BCE and 1900 CE, humans destroyed 2.5 billion acres (1 BN hectares) of forest all over the planet

Over the last century and a bit, we have destroyed another 2.5 billion acres (1 BN hectares) of Earth's forests

Mars volcano Olympus Mons is as big as Arizona" (page 156) and "The largest iceberg reliably recorded was bigger than Corsica and Mallorca" (page 128).

Maps are also time machines. We have gathered several that we hope bring alive the distant past: "The spread of humans and the extinction of large mammals" (pages 4–5), "North America's supervolcanoes" (page 6), and "When the English Channel was a mighty River" (page 7). But in considering the future of the natural world, one subject dominates: the human-caused climate crisis. This book could have comprised 100 maps that all illustrated different aspects of that huge challenge to much of life on the planet. By including maps such as "Who is at risk from rising seas?" (pages 148–49) and "A rapidly warming world" (pages 144–45) to name just two, we hope to leave the reader in little doubt as to the scale of that challenge. But we have also, as Chris Packham notes, tried to find maps—such as "The windy stretches that could help power the world" (pages 76–77) and "The sunny places that could help power the world" (pages 80–81)—that suggest opportunities that might, with the technology, will, and vision, allow us to change our ways.

We hope you enjoy *Wild Maps for Curious Minds*.

MIKE HIGGINS
London, May 2022

1

ANCIENT HISTORY

Empire of the **woolly mammoth**

These Ice Age beasts died out more than 10,000 years ago, except for a few remote herds.

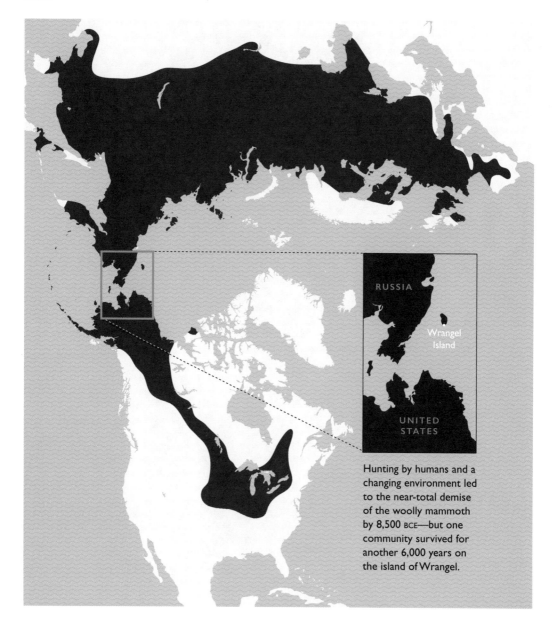

■ Greatest extent of woolly mammoth range

RUSSIA

Wrangel
Island

UNITED
STATES

Hunting by humans and a
changing environment led
to the near-total demise
of the woolly mammoth
by 8,500 BCE—but one
community survived for
another 6,000 years on
the island of Wrangel.

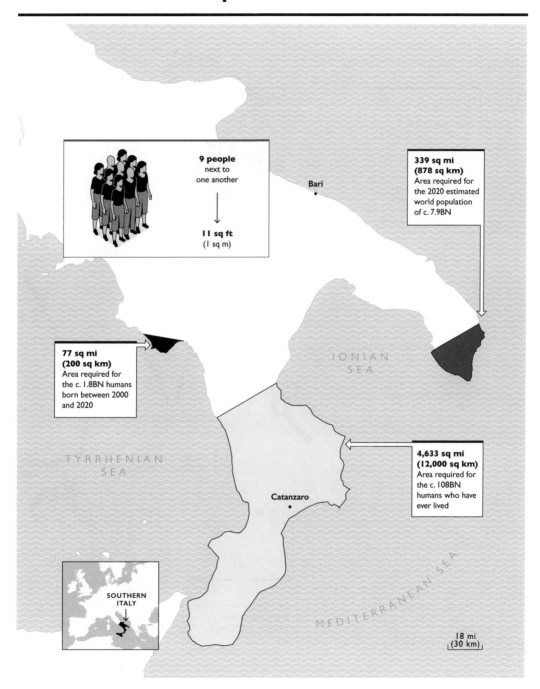

9 people
next to
one another

11 sq ft
(1 sq m)

Bari

339 sq mi
(878 sq km)
Area required for
the 2020 estimated
world population
of c. 7.9BN

77 sq mi
(200 sq km)
Area required for
the c. 1.8BN humans
born between 2000
and 2020

IONIAN
SEA

4,633 sq mi
(12,000 sq km)
Area required for
the c. 108BN
humans who have
ever lived

TYRRHENIAN
SEA

Catanzaro

SOUTHERN
ITALY

MEDITERRANEAN SEA

18 mi
(30 km)

The **spread of humans** and the extinction of **large mammals**

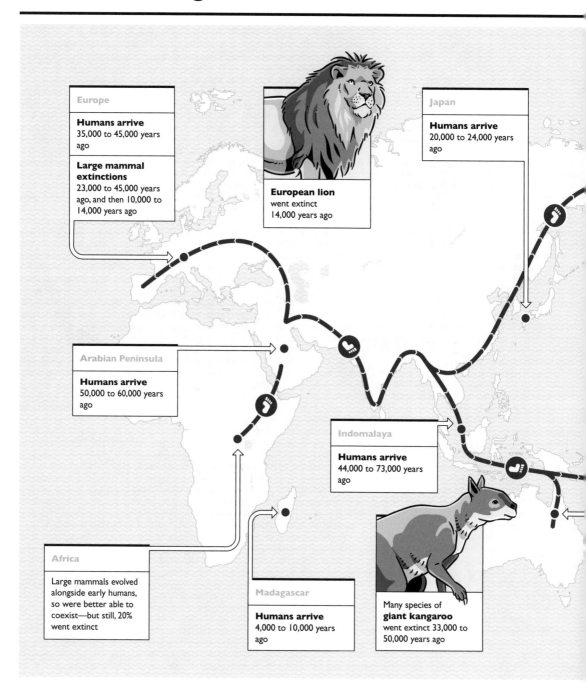

Europe

Humans arrive
35,000 to 45,000 years ago

Large mammal extinctions
23,000 to 45,000 years ago, and then 10,000 to 14,000 years ago

European lion
went extinct
14,000 years ago

Japan

Humans arrive
20,000 to 24,000 years ago

Arabian Peninsula

Humans arrive
50,000 to 60,000 years ago

Indomalaya

Humans arrive
44,000 to 73,000 years ago

Africa

Large mammals evolved alongside early humans, so were better able to coexist—but still, 20% went extinct

Madagascar

Humans arrive
4,000 to 10,000 years ago

Many species of
giant kangaroo
went extinct 33,000 to 50,000 years ago

More than 178 of the world's largest mammals were driven to extinction between 52,000 BCE and 9,000 BCE—at the same time as humans migrated across the world.

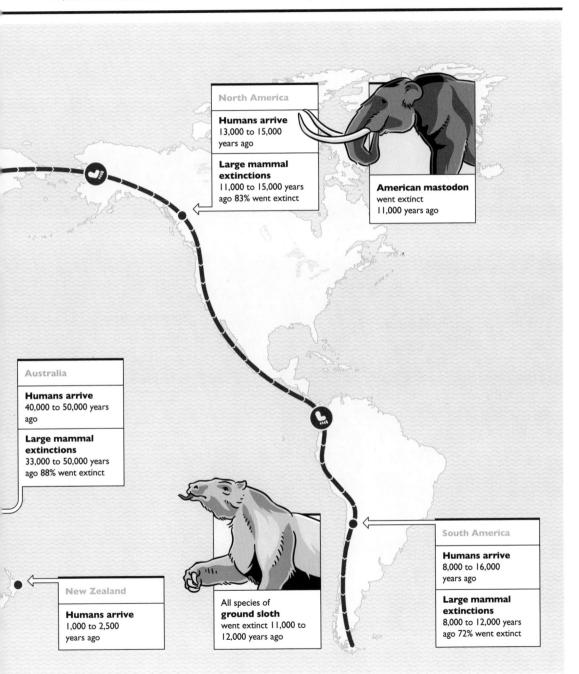

North America

Humans arrive
13,000 to 15,000 years ago

Large mammal extinctions
11,000 to 15,000 years ago 83% went extinct

American mastodon
went extinct
11,000 years ago

Australia

Humans arrive
40,000 to 50,000 years ago

Large mammal extinctions
33,000 to 50,000 years ago 88% went extinct

South America

Humans arrive
8,000 to 16,000 years ago

Large mammal extinctions
8,000 to 12,000 years ago 72% went extinct

All species of **ground sloth** went extinct 11,000 to 12,000 years ago

New Zealand

Humans arrive
1,000 to 2,500 years ago

④ North America's **supervolcanoes**

America's last big volcano eruption in 1980 is dwarfed by four ancient "supereruptions."

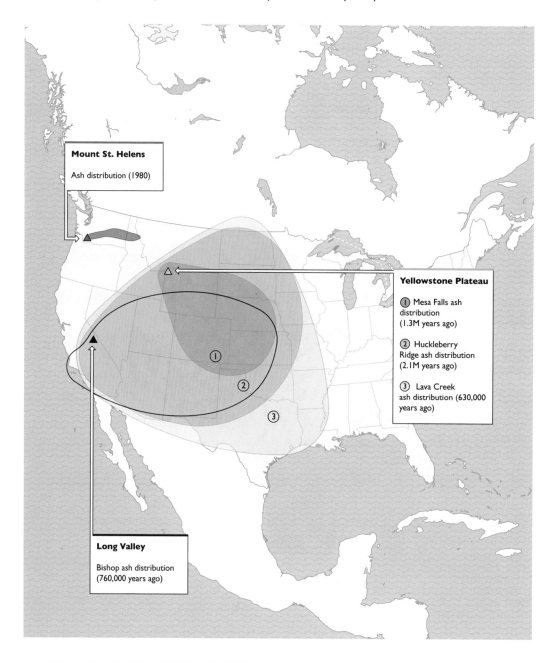

Mount St. Helens

Ash distribution (1980)

Yellowstone Plateau

① Mesa Falls ash distribution (1.3M years ago)

② Huckleberry Ridge ash distribution (2.1M years ago)

③ Lava Creek ash distribution (630,000 years ago)

Long Valley

Bishop ash distribution (760,000 years ago)

5 When the **English Channel** was a mighty river

In the last Ice Age 23,000 years ago, it's thought that many of Europe's rivers flowed into the Fleuve Manche, an ancient river since swamped by the sea.

Fleuve Manche basin

Catchment   Tributaries and rivers   Potential lake   Extent of glacial ice cover

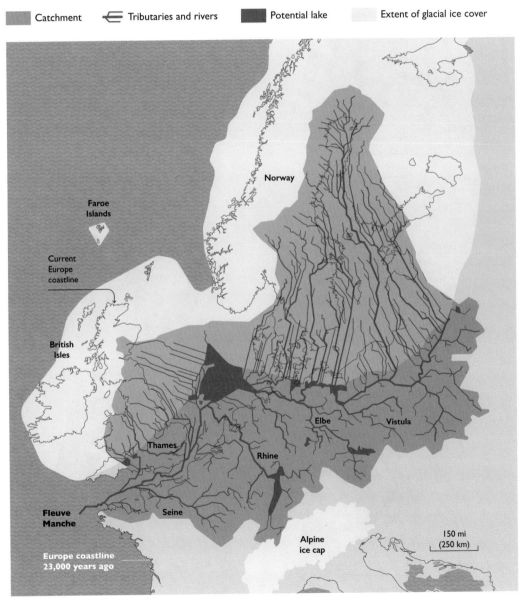

6 The rise and fall of **Europe's forests**

11,000 years ago

10,000 years ago

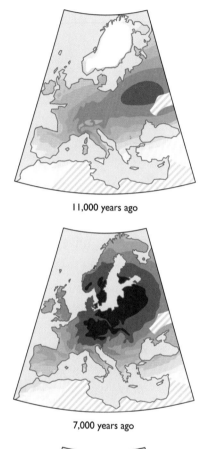

7,000 years ago

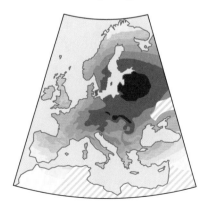

6,000 years ago

3,000 years ago

2,000 years ago

Percentage of land covered by forest

0–10　　10–20　　20–30　　30–40　　40–50　　50–60

60–70　　70–80　　80–90　　90–100　　No data　　Covered by ice

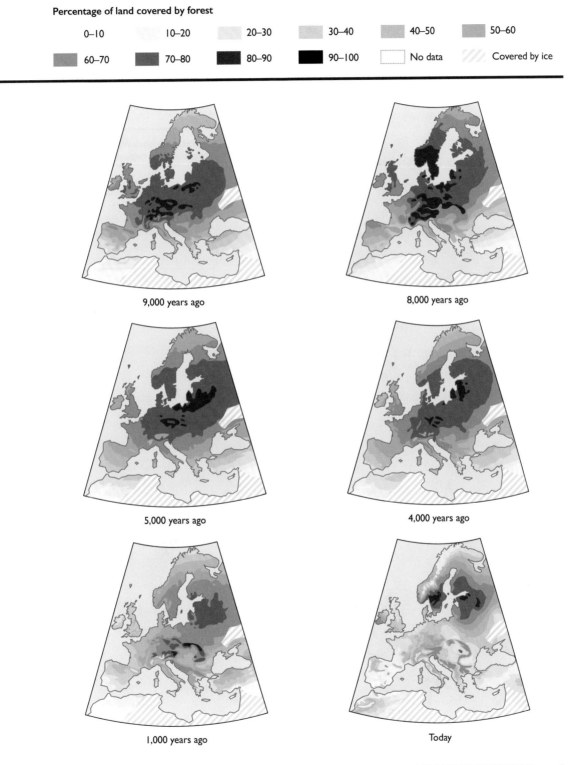

9,000 years ago

8,000 years ago

5,000 years ago

4,000 years ago

1,000 years ago

Today

Where **bears** once ruled in Italy

If you want to see how extensive the ranges of these great carnivores were, just look at how many places are named after them.

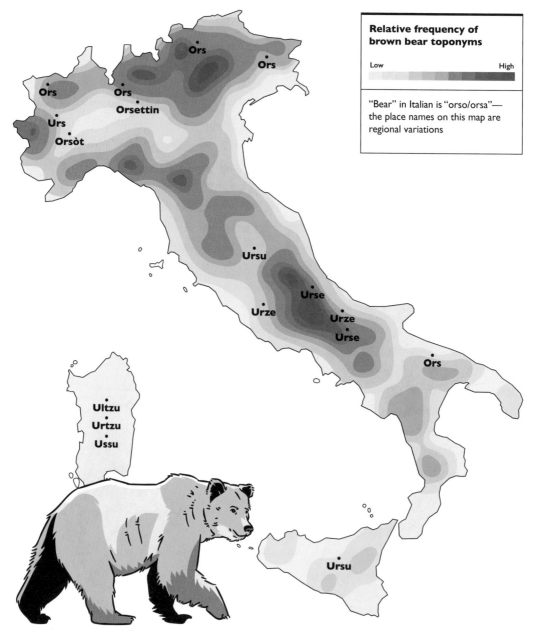

Relative frequency of brown bear toponyms

Low   High

"Bear" in Italian is "orso/orsa"—the place names on this map are regional variations

Ors
Ors
Ors
Ors
Orsettin
Urs
Orsòt
Ursu
Urse
Urze
Urze
Urse
Ors
Ultzu
Urtzu
Ussu
Ursu

Where **wolves** once ruled in Italy

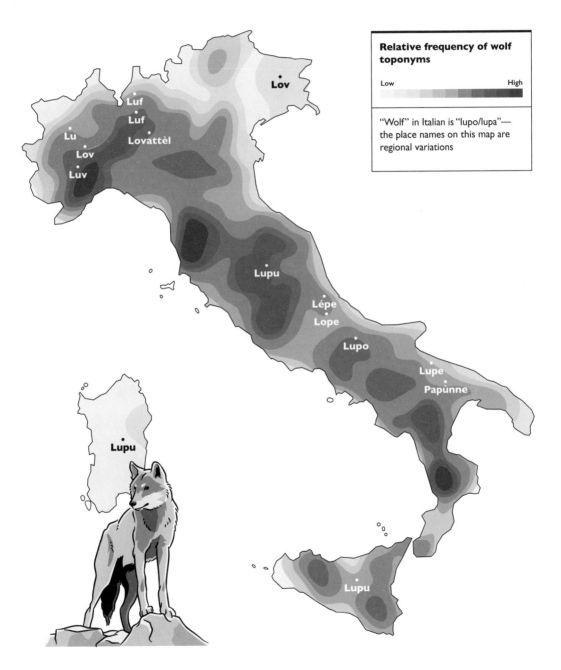

Relative frequency of wolf toponyms

Low | High

"Wolf" in Italian is "lupo/lupa"—the place names on this map are regional variations

Lov

Luf

Luf

Lu

Lovattèl

Lov

Luv

Lupu

Lépe

Lope

Lupo

Lupe

Papùnne

Lupu

Lupu

9 The retreat of the **hunter-gatherer**

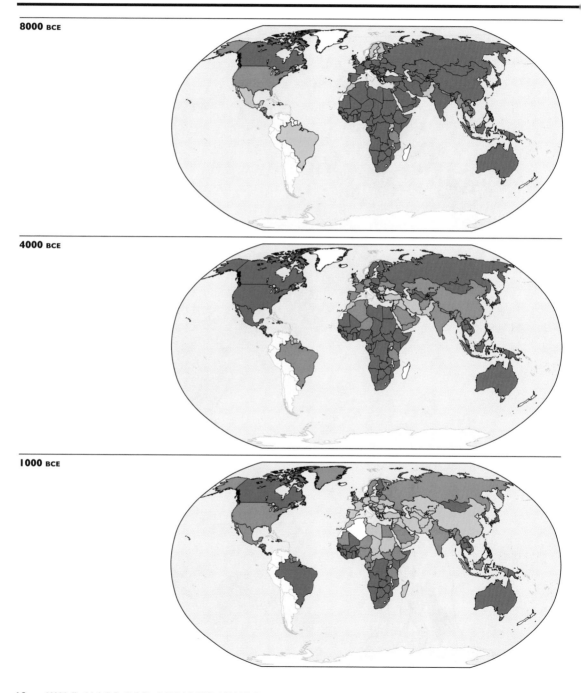

8000 BCE

4000 BCE

1000 BCE

The extent of hunting, gathering, and foraging

| | None | | Minimal (less than 1% of land area) | | Common (1–20%) | | Widespread (more than 20%) | No data |

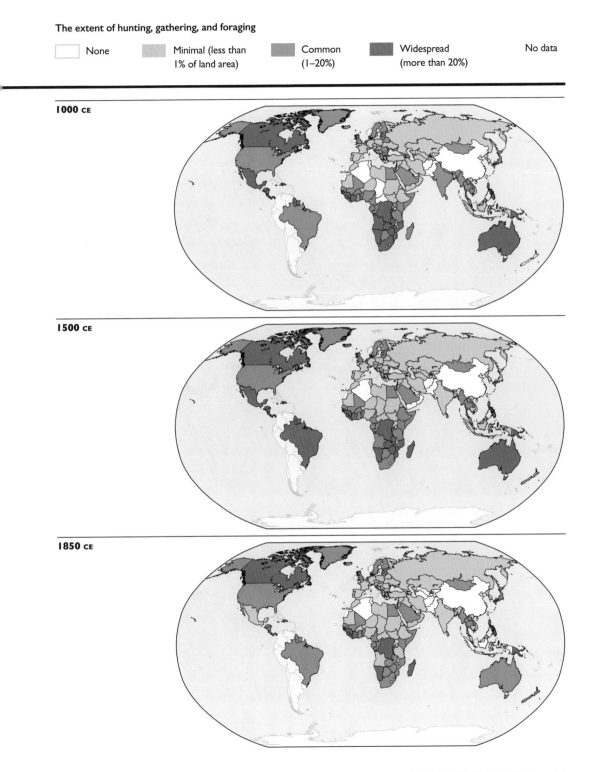

1000 CE

1500 CE

1850 CE

⑩ The advance of the **farmer**

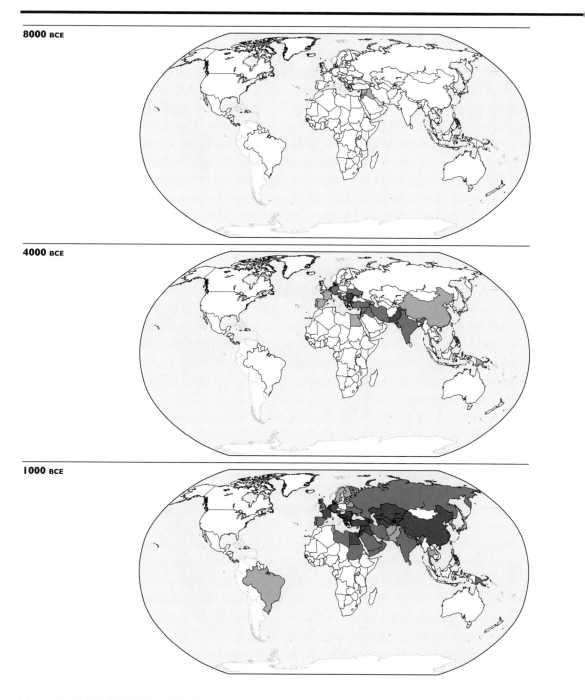

8000 BCE

4000 BCE

1000 BCE

The extent of intensive agriculture

☐ None	Minimal (less than 1% of land area)	Common (1–20%)	Widespread (more than 20%)	No data

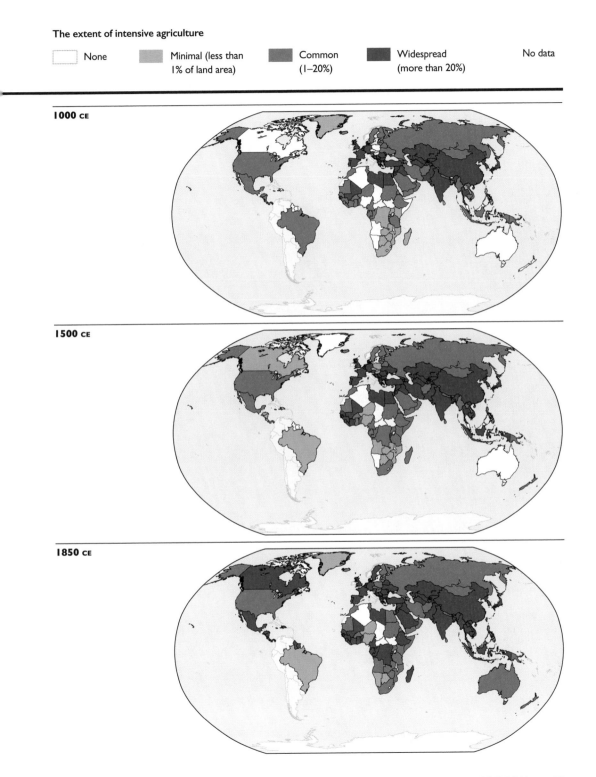

1000 CE

1500 CE

1850 CE

② OUT AND ABOUT

Follow your nose:
The longest walk in a straight line

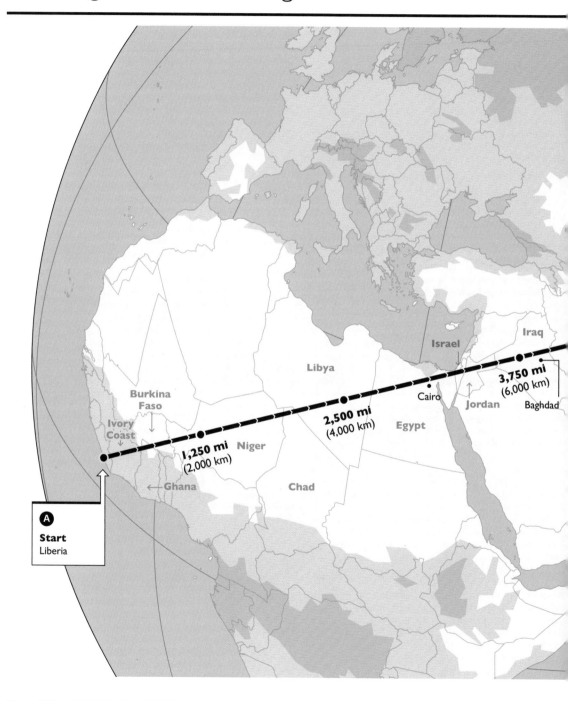

Iraq

Israel

Libya

3,750 mi
(6,000 km)

Burkina
Faso

Cairo

Jordan

Baghdad

Ivory
Coast

2,500 mi
(4,000 km)

Egypt

1,250 mi Niger
(2,000 km)

←Ghana

Chad

A
Start
Liberia

Head northeast-by-east from western Liberia and (in theory) you'll be on the coast of eastern China some 8,400 miles (13,500 km) later.

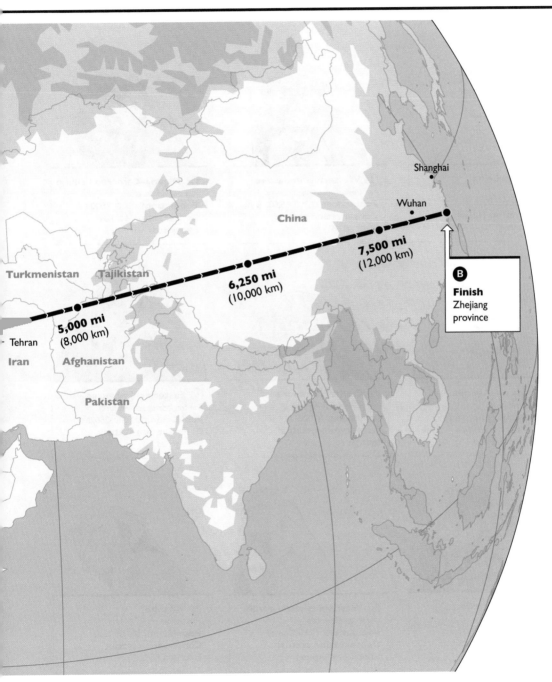

Turkmenistan   Tajikistan

6,250 mi
(10,000 km)

China

Shanghai

Wuhan

7,500 mi
(12,000 km)

5,000 mi
(8,000 km)

Tehran

Iran   Afghanistan

Pakistan

B

Finish
Zhejiang
province

Distribution of the **great apes**

① Human being
Homo sapiens
Population: 7.9BN (est)
Conservation status:
Not evaluated

② Western chimpanzee
Pan troglodytes verus
Population: 21,300–55,600 (est)
Conservation status:
Endangered

③ Nigeria-Cameroon chimp
Pan troglodytes ellioti
Population: 3,500–9,000 (est)
Conservation status:
Endangered

④ Cross river gorilla
Gorilla gorilla diehli
Population: 250–300 (est)
Conservation status:
Critically endangered

⑤ Central chimpanzee
Pan troglodytes troglodytes
Population: 70,000–116,500 (est)
Conservation status:
Endangered

⑥ Eastern chimpanzee
Pan troglodytes schweinfurthii
Population: 200,000–250,000 (est)
Conservation status:
Endangered

⑦ Mountain gorilla
Gorilla beringei beringei
Population: 880 (est)
Conservation status:
Critically endangered

⑧ Western lowland gorilla
Gorilla gorilla gorilla
Population: 150,000 (est)
Conservation status:
Critically endangered

⑨ Bonobo
Pan paniscus
Population: 50,000 (est)
Conservation status:
Endangered

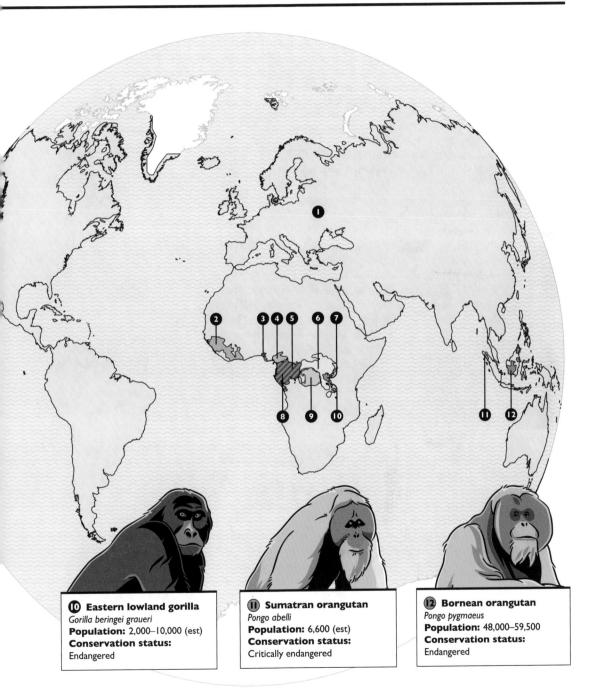

⑩ Eastern lowland gorilla
Gorilla beringei graueri
Population: 2,000–10,000 (est)
Conservation status:
Endangered

⑪ Sumatran orangutan
Pongo abelli
Population: 6,600 (est)
Conservation status:
Critically endangered

⑫ Bornean orangutan
Pongo pygmaeus
Population: 48,000–59,500
Conservation status:
Endangered

13 Not every **giraffe** looks alike

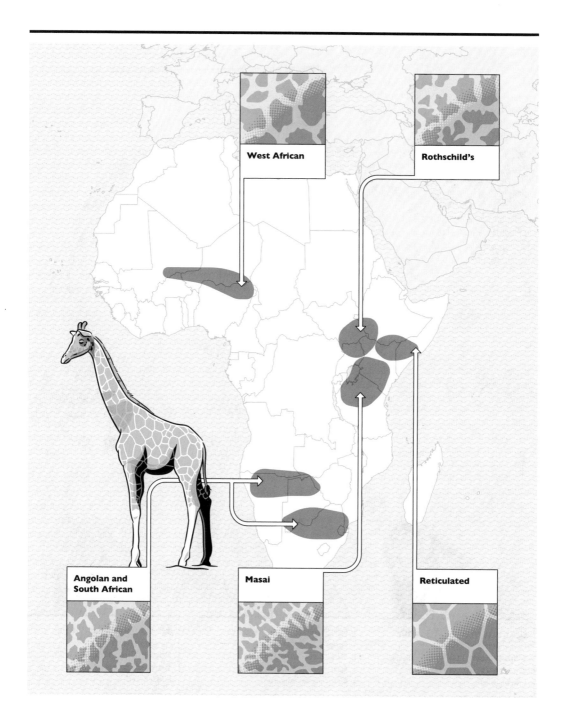

West African

Rothschild's

Angolan and
South African

Masai

Reticulated

Where Americans get their wilderness:
The top 10 most visited US National Parks

Americans love their national parks; the 10 parks below received nearly 50 million visitors combined in 2019.

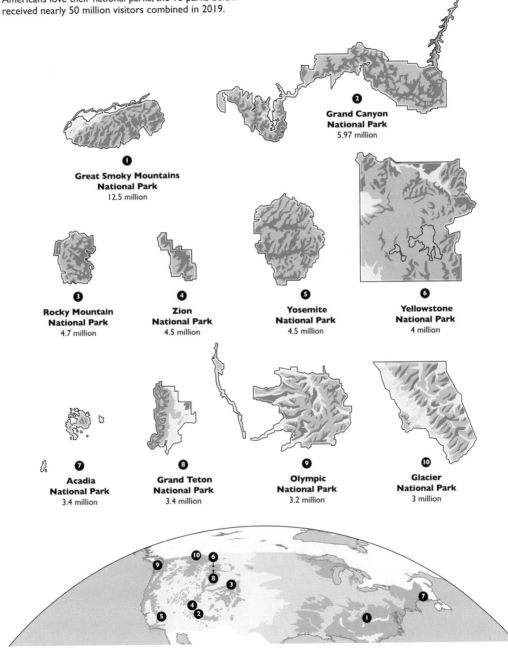

②
Grand Canyon National Park
5.97 million

①
Great Smoky Mountains National Park
12.5 million

③
Rocky Mountain National Park
4.7 million

④
Zion National Park
4.5 million

⑤
Yosemite National Park
4.5 million

⑥
Yellowstone National Park
4 million

⑦
Acadia National Park
3.4 million

⑧
Grand Teton National Park
3.4 million

⑨
Olympic National Park
3.2 million

⑩
Glacier National Park
3 million

Where **cattle, sheep, or pigs** outnumber people

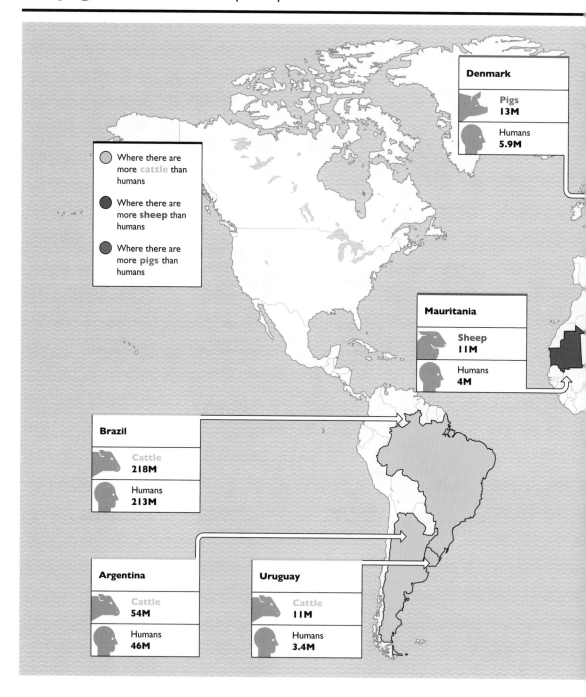

Denmark

| Pigs | 13M |
| Humans | 5.9M |

Where there are more cattle than humans

Where there are more sheep than humans

Where there are more pigs than humans

Mauritania

| Sheep | 11M |
| Humans | 4M |

Brazil

| Cattle | 218M |
| Humans | 213M |

Argentina

| Cattle | 54M |
| Humans | 46M |

Uruguay

| Cattle | 11M |
| Humans | 3.4M |

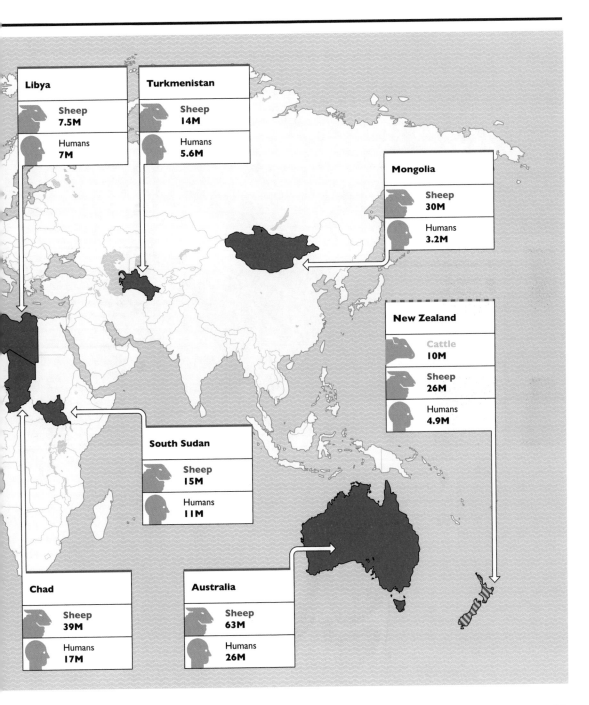

Libya

| Sheep | 7.5M |
| Humans | 7M |

Turkmenistan

| Sheep | 14M |
| Humans | 5.6M |

Mongolia

| Sheep | 30M |
| Humans | 3.2M |

New Zealand

Cattle	10M
Sheep	26M
Humans	4.9M

South Sudan

| Sheep | 15M |
| Humans | 11M |

Chad

| Sheep | 39M |
| Humans | 17M |

Australia

| Sheep | 63M |
| Humans | 26M |

Where to see a **giant panda** outside China

Copenhagen Zoo
Denmark

Xing Er
and Mao Sun

Edinburgh Zoo
UK

Yang Guang
and Tian Tian

Smithsonian's National Zoological Park United States

Tian Tian, Mei Xiang, and Xiao Qi Ji

Ouwehands Dierenpark
Netherlands

Xing Ya, Wu Wen, and Fan Xing

Memphis Zoo
United States

Le Le and Ya Ya

Madrid Zoo
Spain

Bing Xing, Hua Zui Ba, Jiu Jiu, and You You

Zoo Atlanta
United States

Ya Lun, Xi Lun, and Yang Yang

Zooparc Beauval
France

Huan Huan, Yuan Zai, Huan Lili, and Yuan Dudu

Chapultepec Zoo
Mexico

Shuan Shuan and Xin Xin

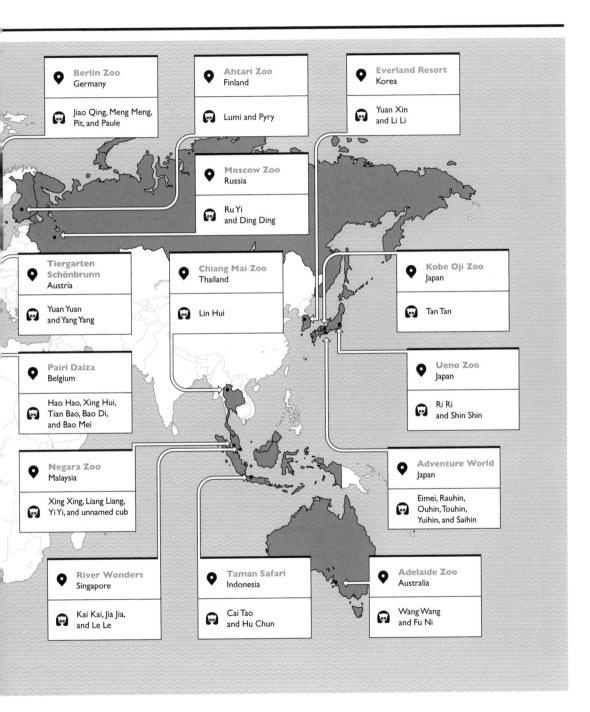

Berlin Zoo
Germany

Jiao Qing, Meng Meng, Pit, and Paule

Ahtari Zoo
Finland

Lumi and Pyry

Everland Resort
Korea

Yuan Xin and Li Li

Moscow Zoo
Russia

Ru Yi and Ding Ding

Tiergarten Schönbrunn
Austria

Yuan Yuan and Yang Yang

Chiang Mai Zoo
Thailand

Lin Hui

Kobe Oji Zoo
Japan

Tan Tan

Pairi Daiza
Belgium

Hao Hao, Xing Hui, Tian Bao, Bao Di, and Bao Mei

Ueno Zoo
Japan

Ri Ri and Shin Shin

Negara Zoo
Malaysia

Xing Xing, Liang Liang, Yi Yi, and unnamed cub

Adventure World
Japan

Eimei, Rauhin, Ouhin, Touhin, Yuihin, and Saihin

River Wonders
Singapore

Kai Kai, Jia Jia, and Le Le

Taman Safari
Indonesia

Cai Tao and Hu Chun

Adelaide Zoo
Australia

Wang Wang and Fu Ni

17 Which Americans go on **wildlife-watching** adventures?

Americans who travel to see their wildlife at least once a year (percentage of population)

1–4	5–8	9–12	13–16	17–20

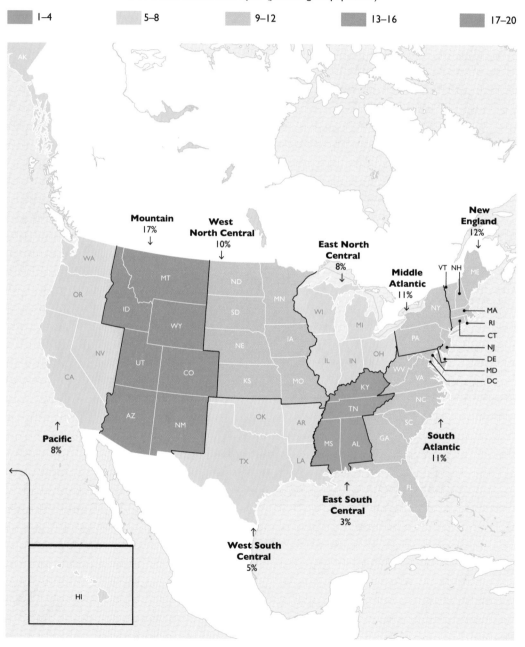

Mountain
17%
↓

West
North Central
10%
↓

East North
Central
8%
↓

New
England
12%
↓

Middle
Atlantic
11%
↓

Pacific
8%
↑

South
Atlantic
11%
↑

East South
Central
3%
↑

West South
Central
5%
↑

18 All the **private gardens** in the UK

The UK's residential gardens, if put together, would form a green space bigger than the largest national park and more than three times the area of London.

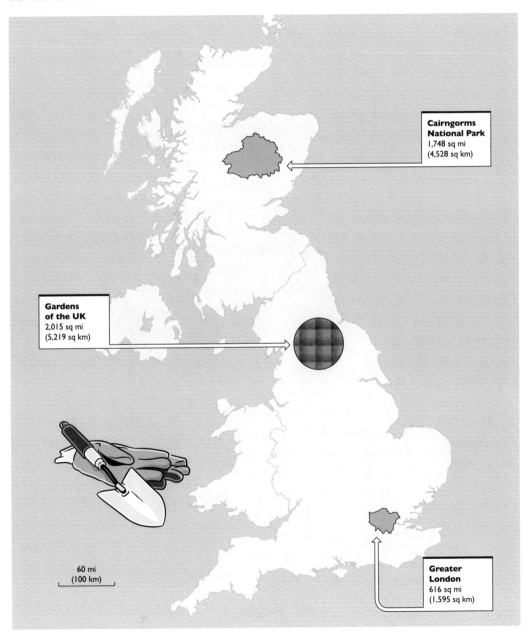

Cairngorms National Park
1,748 sq mi
(4,528 sq km)

Gardens of the UK
2,015 sq mi
(5,219 sq km)

Greater London
616 sq mi
(1,595 sq km)

60 mi
(100 km)

Where to live if you **hate snakes**

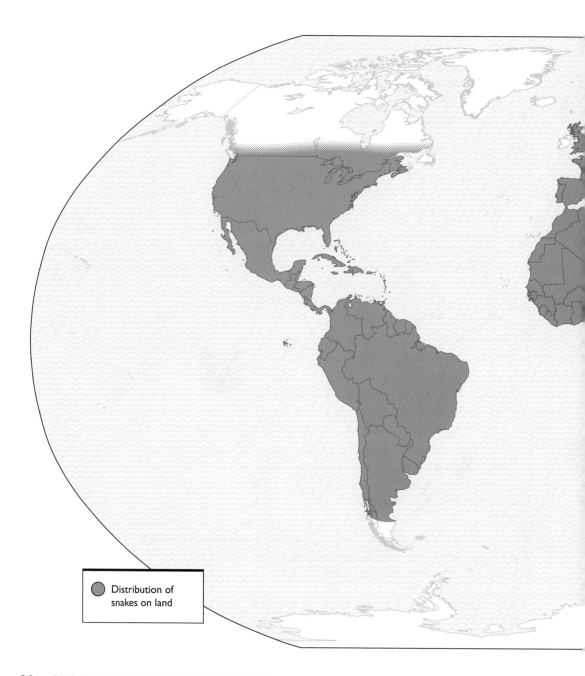

Distribution of snakes on land

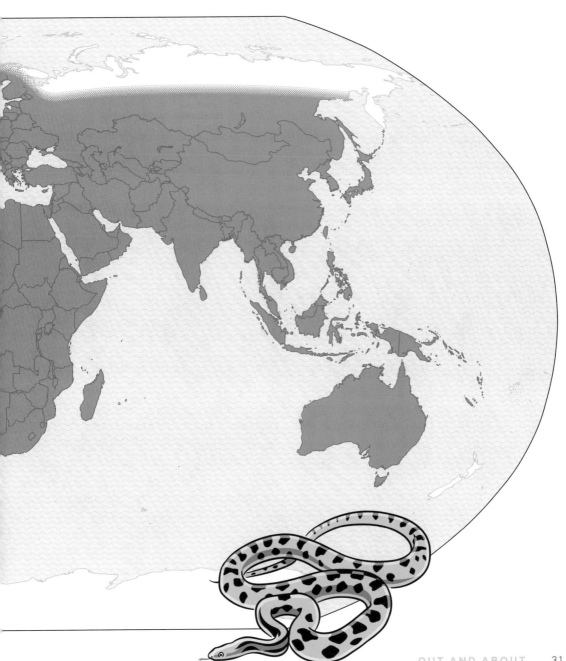

20 **Spoiler alert!** Captain Ahab's fatal pursuit of Moby Dick, the Great White Whale

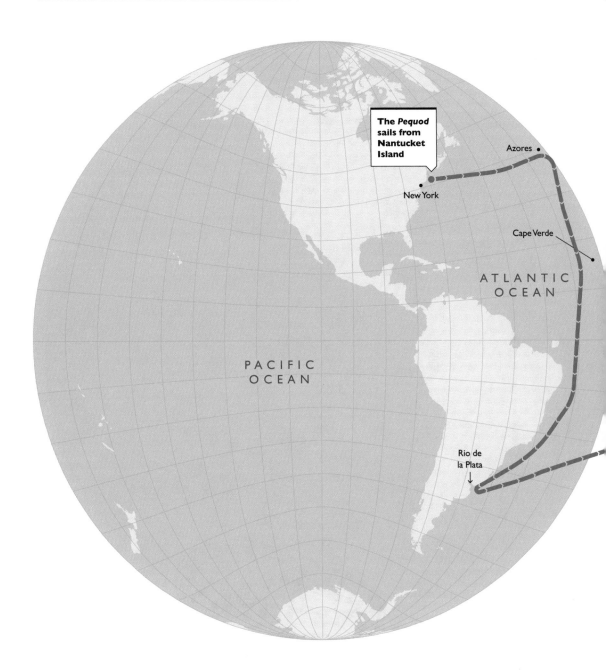

The *Pequod* sails from Nantucket Island

Azores

New York

Cape Verde

ATLANTIC OCEAN

PACIFIC OCEAN

Rio de la Plata ↓

This is the voyage of the whaling ship *Pequod* from Herman Melville's 1851 novel.

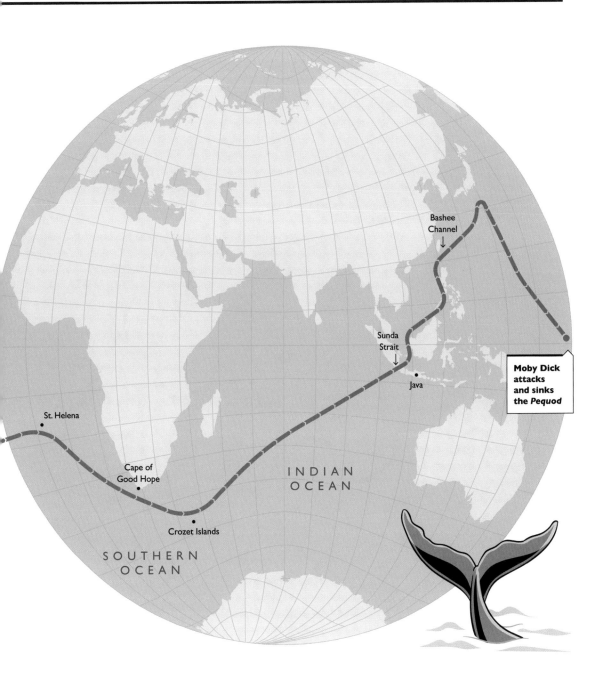

Bashee
Channel

Sunda
Strait

Java

**Moby Dick
attacks
and sinks
the *Pequod***

St. Helena

Cape of
Good Hope

INDIAN
OCEAN

Crozet Islands

SOUTHERN
OCEAN

The **secret rivers** of London

Beneath the streets and buildings of the capital flow half-forgotten tributaries of the Thames.

—————— Open rivers   - - - - - Underground rivers

Salmon's Brook

SOUTHGATE

EDMONTON

Pymme's Brook

EDGWARE   FINCHLEY

MUSWELL
HILL

Silk
Stream   HENDON   TOTTENHAM   River
Ching

HARROW   Mutton Brook

Dollis Brook   Moselle

Wealdstone
Brook   WALTHAMSTOW

HIGHGATE   STOKE NEWINGTON

WEMBLEY   NEASDEN   River
Lea

HAMPSTEAD   Hackney
Brook

WILLESDEN
Westbourne   River Fleet   ISLINGTON   HACKNEY

River Brent   MARYLEBONE

Tyburn   Walbrook

ACTON   Counter's   PADDINGTON   CITY   STEPNEY
Creek
EALING   Stamford
Brook   KENSINGTON   WESTMINSTER

HAMMERSMITH   Neckinger

Thames
KEW   CHELSEA   CAMBERWELL   GREENWICH
Earl's Sluice

BATTERSEA
BRIXTON
Beverley Brook   Falcon
WANDSWORTH   Effra River   Ravensbourne

LEWISHAM
RICHMOND   BALHAM
WIMBLEDON   NORWOOD
STREATHAM
CRYSTAL
PALACE
KINGSTON   River Wandle   River
Graveney
Pyl Brook
MITCHAM   BECKENHAM
Hogsmill River   MORDEN   1.2 mi
CROYDON   Pool River   (2 km)

㉒ How **fast** are you spinning?

As you stand on Earth, your rotational speed depends on how close you are to its axis. In other words, the closer you are to the Equator, the faster you're spinning.

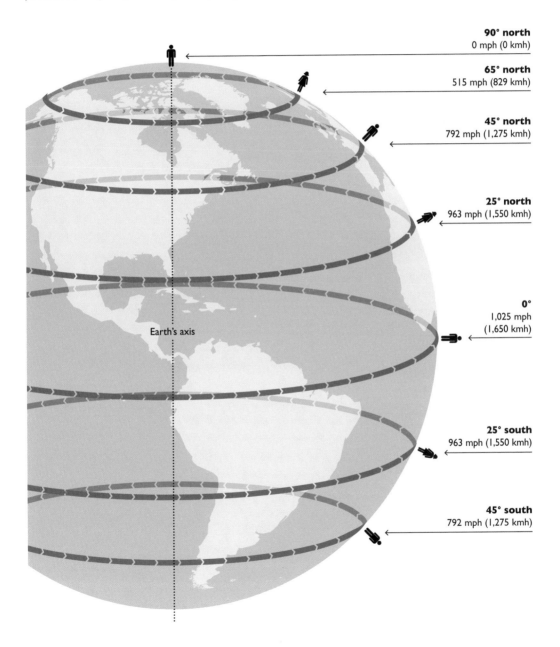

90° north
0 mph (0 kmh)

65° north
515 mph (829 kmh)

45° north
792 mph (1,275 kmh)

25° north
963 mph (1,550 kmh)

0°
1,025 mph
(1,650 kmh)

25° south
963 mph (1,550 kmh)

45° south
792 mph (1,275 kmh)

Earth's axis

③ THE WATERY WORLD

23 Imagine all the oceans as a **single body of water**

The oceanographer Athelstan Spilhaus did just that in his 1942 map projection.

Warm currents

Cool currents

ARCTIC   SIBERIA

EASTERN
NORTH
AMERICA

EUROPE

SOUTHEAST
ASIA

NORTH
ATLANTIC

AFRICA

INDIAN

EASTERN
SOUTH
AMERICA

SOUTH
ATLANTIC

AUSTRALIA

ANTARCTICA

SOUTH
PACIFIC

NORTH
PACIFIC

WESTERN
NORTH
AMERICA

WESTERN
SOUTH
AMERICA

The **tiny creek** that connects
the Atlantic and Pacific

At Two Ocean Creek, the waters of a small stream part and flow either 1,250 miles (2,000 km) west or 1,350 miles (5,500 km) southeast.

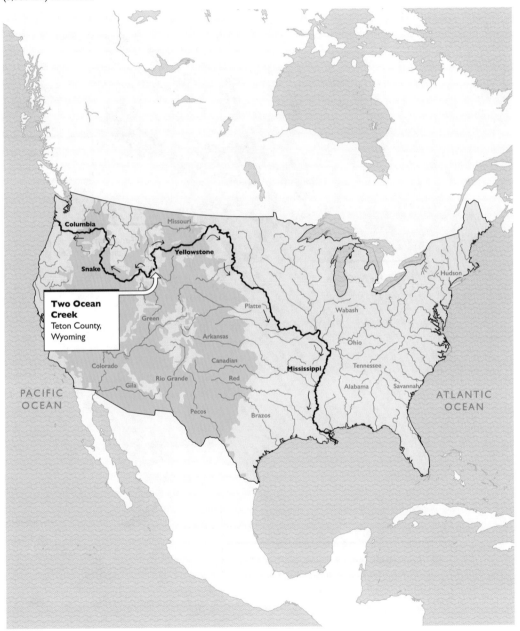

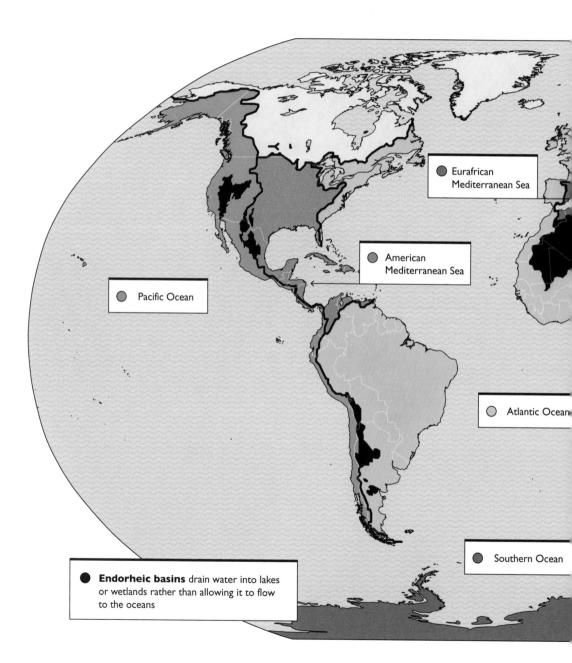

Eurafrican
Mediterranean Sea

American
Mediterranean Sea

Pacific Ocean

Atlantic Ocean

Southern Ocean

Endorheic basins drain water into lakes
or wetlands rather than allowing it to flow
to the oceans

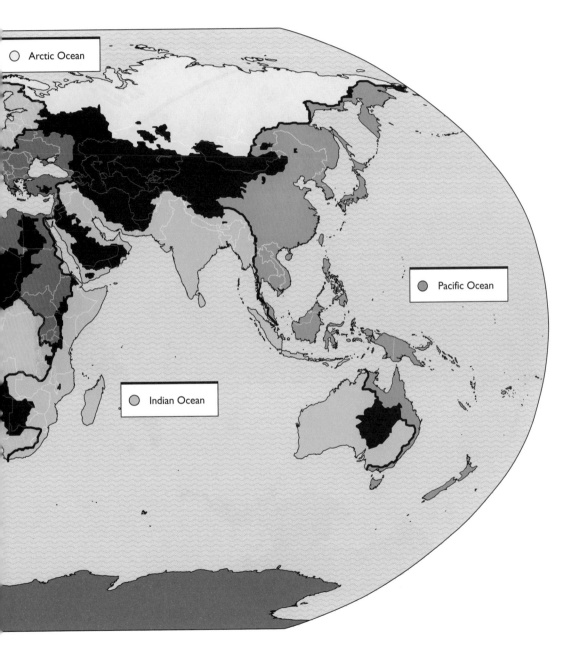

Arctic Ocean

Pacific Ocean

Indian Ocean

㉖ Point Nemo:
The most remote place on Earth

It's just a spot in the Pacific Ocean more than 1,550 miles (2,500 km) from the nearest land—which is why it has become a spacecraft junkyard.

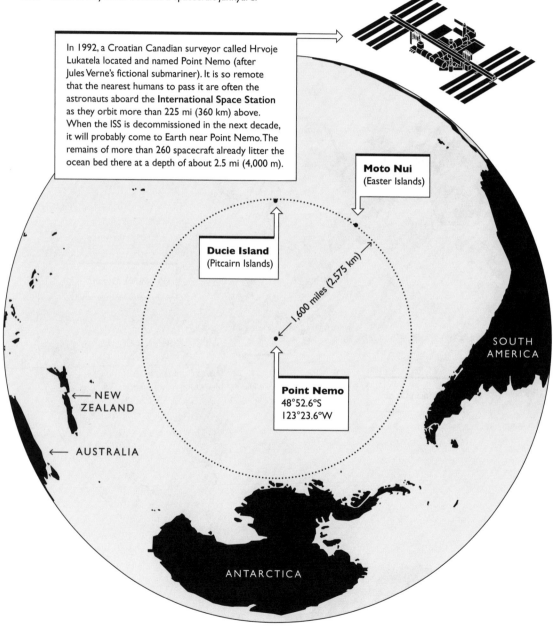

In 1992, a Croatian Canadian surveyor called Hrvoje Lukatela located and named Point Nemo (after Jules Verne's fictional submariner). It is so remote that the nearest humans to pass it are often the astronauts aboard the **International Space Station** as they orbit more than 225 mi (360 km) above. When the ISS is decommissioned in the next decade, it will probably come to Earth near Point Nemo. The remains of more than 260 spacecraft already litter the ocean bed there at a depth of about 2.5 mi (4,000 m).

Moto Nui
(Easter Islands)

Ducie Island
(Pitcairn Islands)

1,600 miles (2,575 km)

Point Nemo
48°52.6°S
123°23.6°W

SOUTH
AMERICA

← NEW
ZEALAND

← AUSTRALIA

ANTARCTICA

Lighthouses of Britain

27

Mariners navigating around Britain are aided by more than 200 lighthouses. Seven have a range of 30 miles (48 km) or farther.

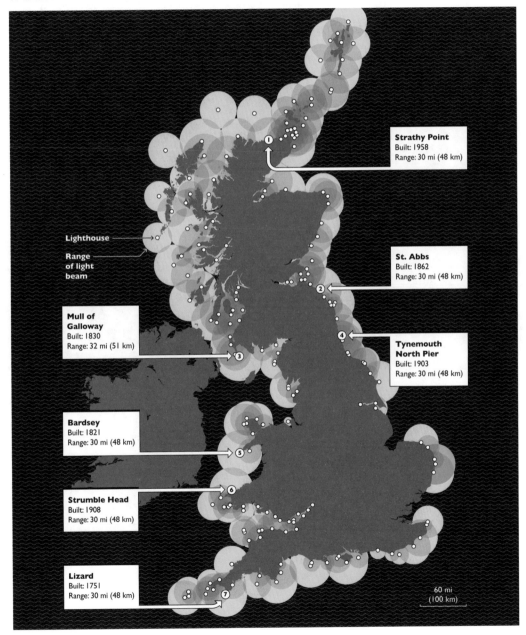

Strathy Point
Built: 1958
Range: 30 mi (48 km)

St. Abbs
Built: 1862
Range: 30 mi (48 km)

Lighthouse

Range
of light
beam

**Tynemouth
North Pier**
Built: 1903
Range: 30 mi (48 km)

**Mull of
Galloway**
Built: 1830
Range: 32 mi (51 km)

Bardsey
Built: 1821
Range: 30 mi (48 km)

Strumble Head
Built: 1908
Range: 30 mi (48 km)

Lizard
Built: 1751
Range: 30 mi (48 km)

60 mi
(100 km)

Annual average rainfall, in inches

- 0–10 (0–25 cm)
- 10–20 (25–50 cm)
- 20–30 (50–75 cm)
- 30–40 (75–100 cm)
- 40–50 (100–125 cm)
- 50–60 (125–150 cm)
- 60–70 (150–175 cm)
- More than 70 (175 cm)

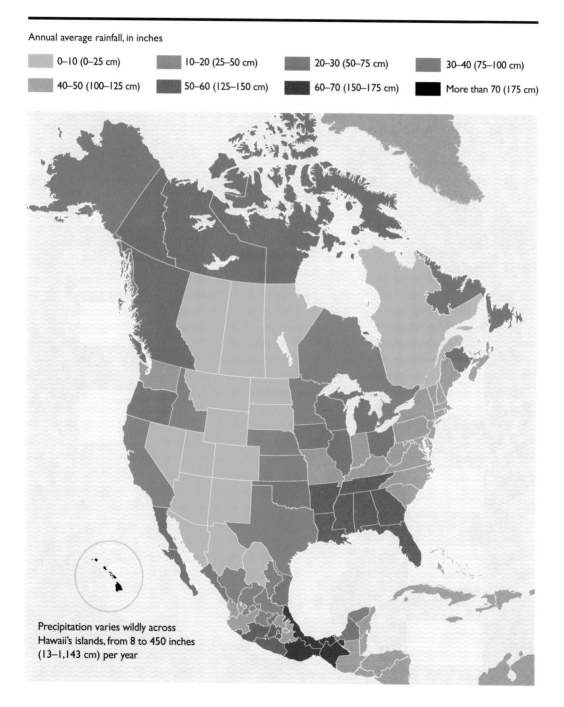

Precipitation varies wildly across
Hawaii's islands, from 8 to 450 inches
(13–1,143 cm) per year

29 Which is the **wettest** half of the year in North America?

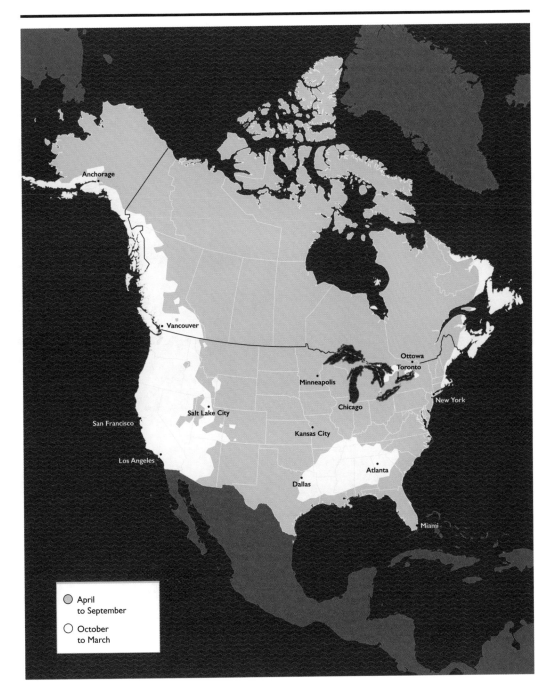

Anchorage

Vancouver

Ottawa
Toronto

Minneapolis

Chicago

New York

Salt Lake City

San Francisco

Kansas City

Los Angeles

Atlanta

Dallas

Miami

- April to September
- October to March

30

It's raining, but is it **pouring?**

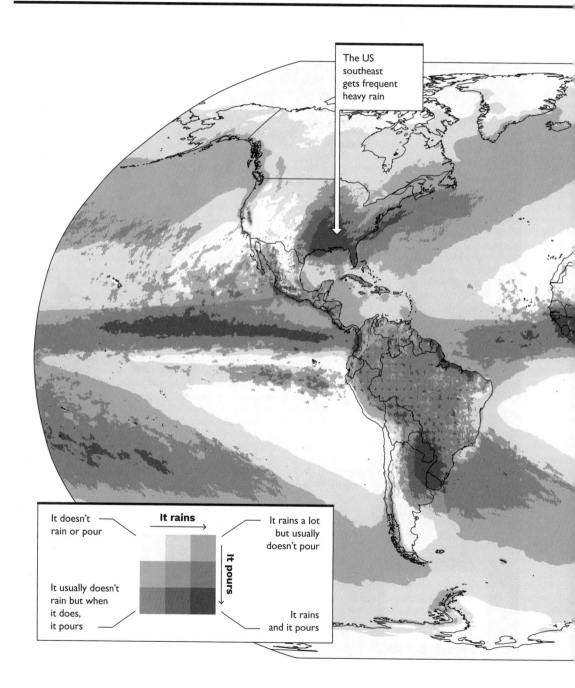

The US southeast gets frequent heavy rain

It doesn't rain or pour

It rains →

It rains a lot but usually doesn't pour

It pours ↓

It usually doesn't rain but when it does, it pours

It rains and it pours

Certain places get frequent light rain, others get occasional downpours—
and in some spots the heavy rain just keeps coming.

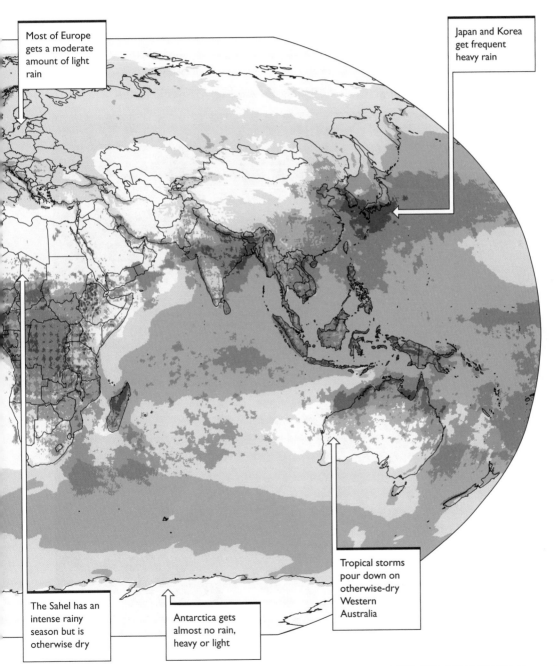

Most of Europe gets a moderate amount of light rain

Japan and Korea get frequent heavy rain

The Sahel has an intense rainy season but is otherwise dry

Antarctica gets almost no rain, heavy or light

Tropical storms pour down on otherwise-dry Western Australia

31

Level of **water stress** across North America

"Water stress" represents how likely it is that a given location might run out of water in a typical year. But "typical" years are evolving with the increased strain on renewable surface and groundwater supplies caused by climate change. A dry spell in New Mexico, the most water-stressed US state, could put it over the edge; currently its stress level is on par with the United Arab Emirates. Mexico City, another place that falls within the "extremely high" category, is currently drawing so much of its groundwater that it's sinking.

Water stress level Low Low–medium Medium–high High
Extremely high Arid/low water use No data

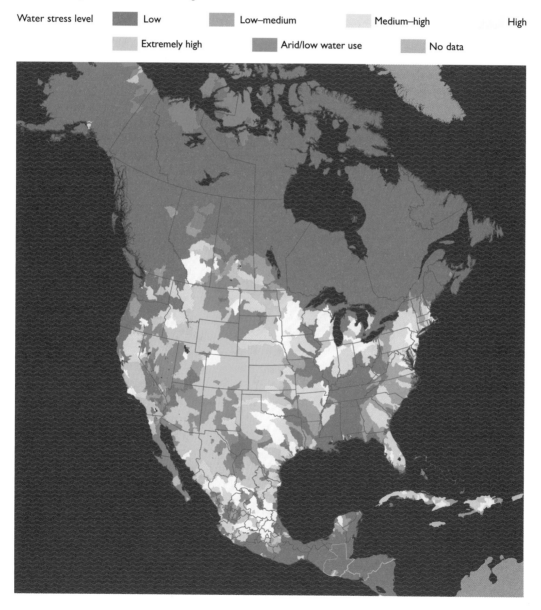

32 Where it might snow on **Christmas Day** across North America

Probability of measurable new snow on Christmas Day (40-year historical period)

Over 35%   30–35%   25–30%   20–25%

15–20%   10–15%   5–10%   0.5–5%

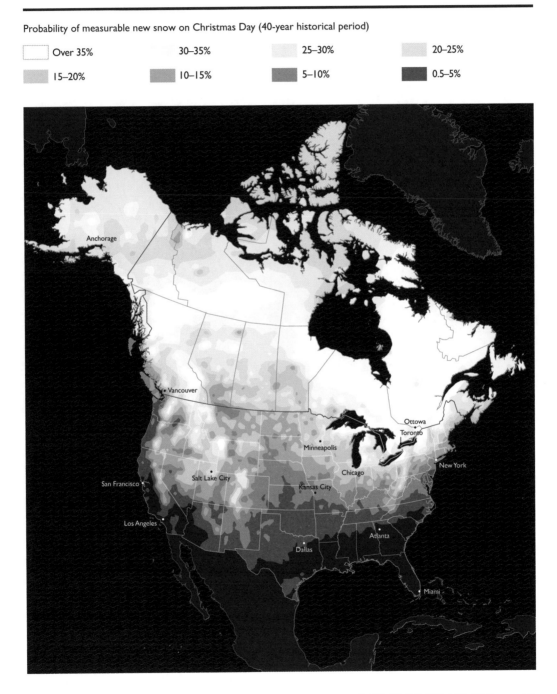

4

GEOGRAPHY

33 These spheres represent all the **water and air** on Earth

This map may be hard to believe, but bear in mind that, while the oceans may be deep (nearly 7 miles/11 km at their deepest), the earth is far deeper, at nearly 4,000 miles/6,400 km from surface to core (see next page).

Water
All the world's water would form a sphere just under 900 mi (1,400 km) in diameter (97% of which would be seawater)

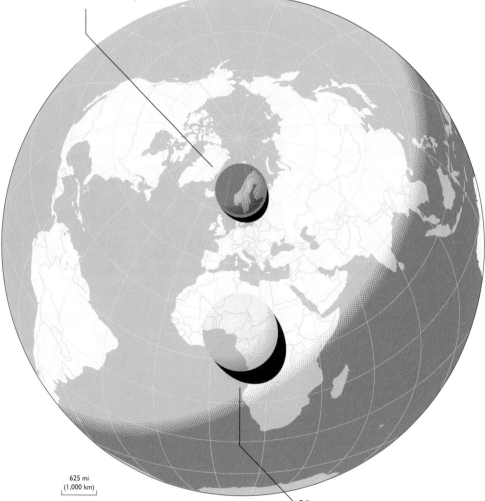

625 mi
(1,000 km)

Air
All the air we breathe could be contained in a sphere just 1,250 mi (2,000 km) in diameter, at standard pressure and 59°F (15°C)

34 How **deep** is the Earth?

The distance from the top of North America to the bottom is about the same as the distance from the Earth's crust to its core.

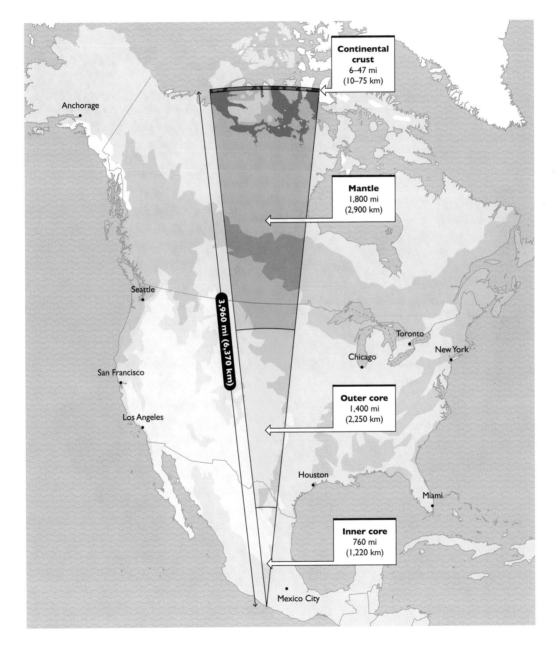

Continental crust
6–47 mi
(10–75 km)

Mantle
1,800 mi
(2,900 km)

3,960 mi (6,370 km)

Outer core
1,400 mi
(2,250 km)

Inner core
760 mi
(1,220 km)

Anchorage

Seattle

San Francisco

Los Angeles

Toronto

New York

Chicago

Houston

Miami

Mexico City

35

The middles of **nowhere**

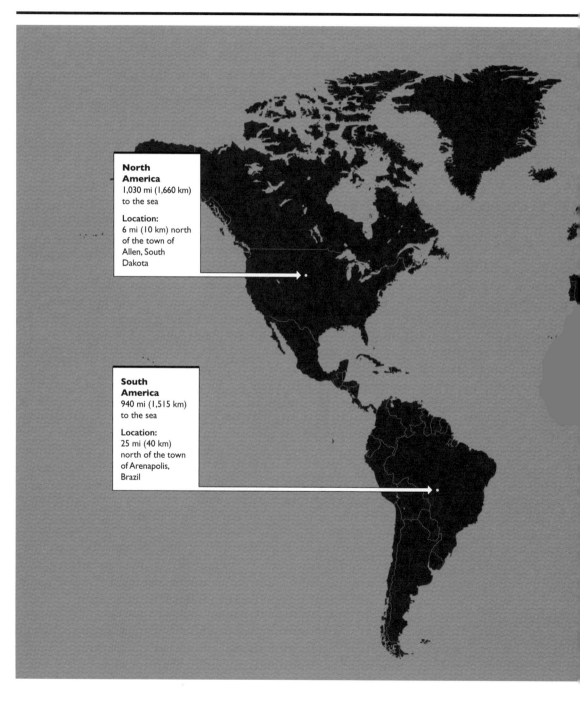

North America
1,030 mi (1,660 km) to the sea

Location:
6 mi (10 km) north of the town of Allen, South Dakota

South America
940 mi (1,515 km) to the sea

Location:
25 mi (40 km) north of the town of Arenapolis, Brazil

The most distant places from the ocean, also known as the continental poles of inaccessibility.

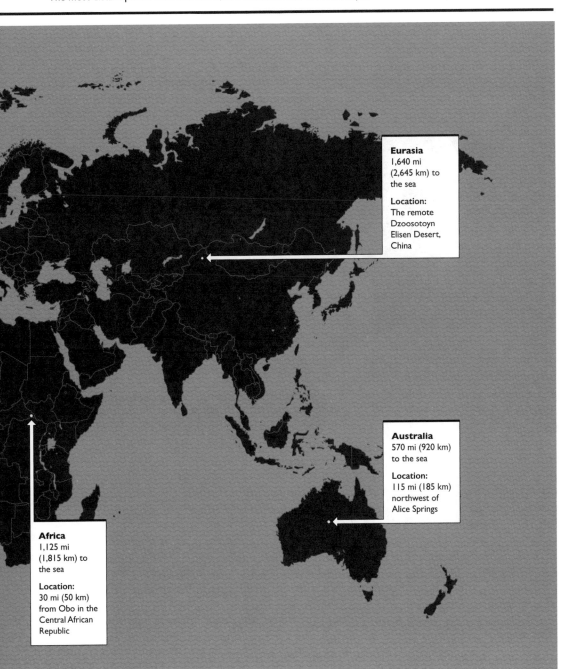

Eurasia
1,640 mi (2,645 km) to the sea

Location: The remote Dzoosotoyn Elisen Desert, China

Australia
570 mi (920 km) to the sea

Location: 115 mi (185 km) northwest of Alice Springs

Africa
1,125 mi (1,815 km) to the sea

Location: 30 mi (50 km) from Obo in the Central African Republic

Which countries lie directly
east and **west** of the Americas?

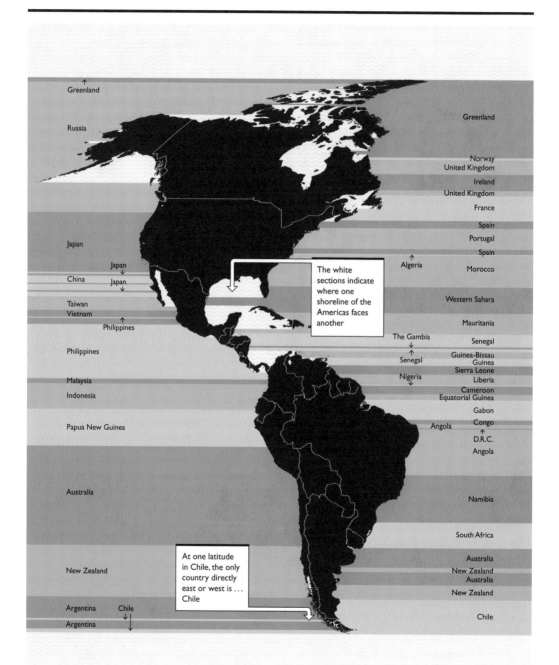

The white sections indicate where one shoreline of the Americas faces another

At one latitude in Chile, the only country directly east or west is ... Chile

What **Antarctica** looks like beneath the ice

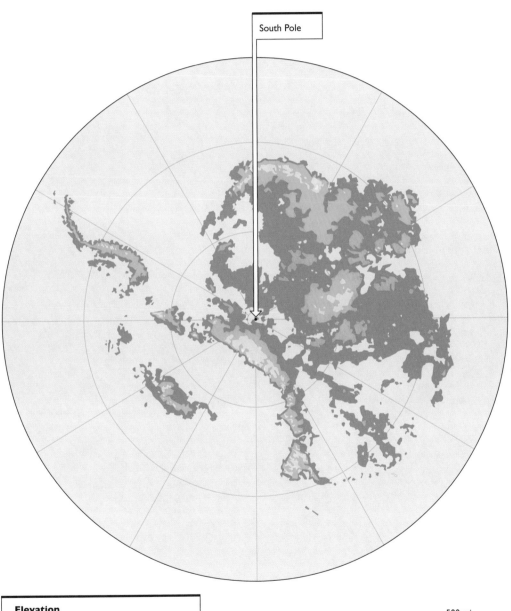

South Pole

Elevation

0 ft	5,000 ft	10,000 ft
(0 m)	(1,500 m)	(3,050 m)

500 mi
(805 km)

The nations with **no sea view**

Czech Republic

Austria

Liechtenstein

Luxembourg

Switzerland

Andorra

San Marino

Vatican City

Niger

Mali

Burkina Faso

Central African Republic

Bolivia

Paraguay

Zambia

Botswana

○ Landlocked countries

● Double-landlocked countries

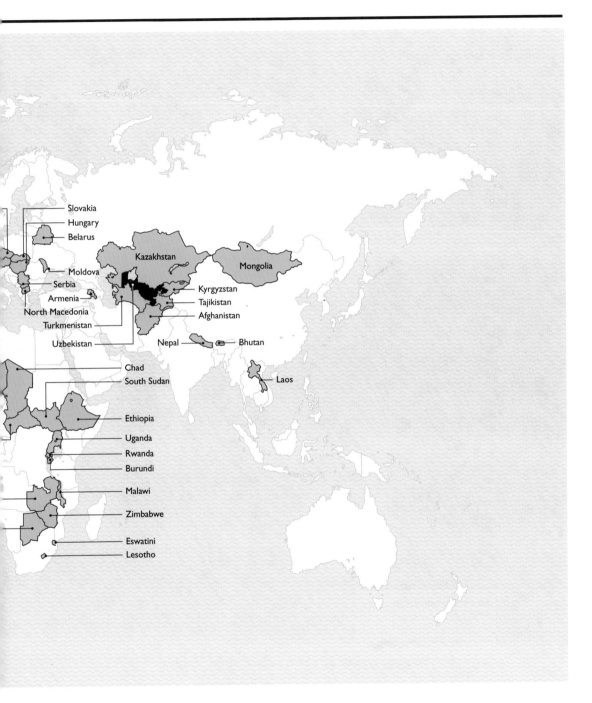

Slovakia
Hungary
Belarus
Kazakhstan
Mongolia
Moldova
Serbia
Kyrgyzstan
Armenia
Tajikistan
North Macedonia
Afghanistan
Turkmenistan
Uzbekistan
Nepal
Bhutan
Chad
South Sudan
Laos
Ethiopia
Uganda
Rwanda
Burundi
Malawi
Zimbabwe
Eswatini
Lesotho

39

All the **rivers** in the world

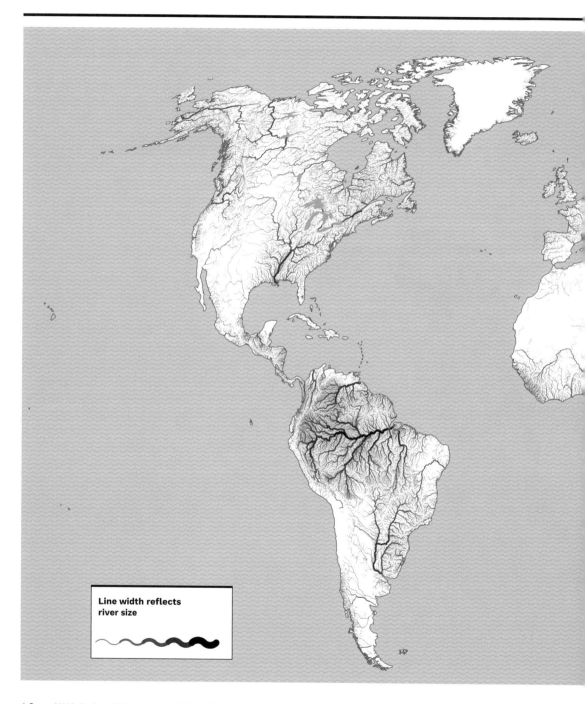

Line width reflects river size

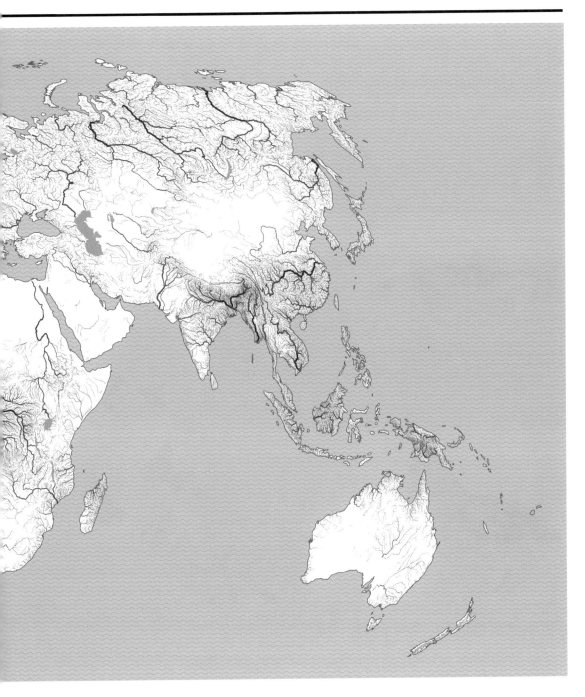

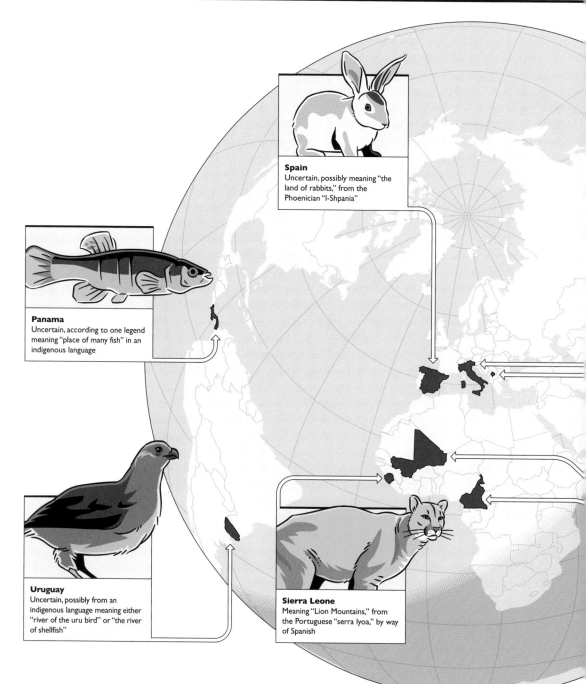

Spain
Uncertain, possibly meaning "the land of rabbits," from the Phoenician "I-Shpania"

Panama
Uncertain, according to one legend meaning "place of many fish" in an indigenous language

Uruguay
Uncertain, possibly from an indigenous language meaning either "river of the uru bird" or "the river of shellfish"

Sierra Leone
Meaning "Lion Mountains," from the Portuguese "serra lyoa," by way of Spanish

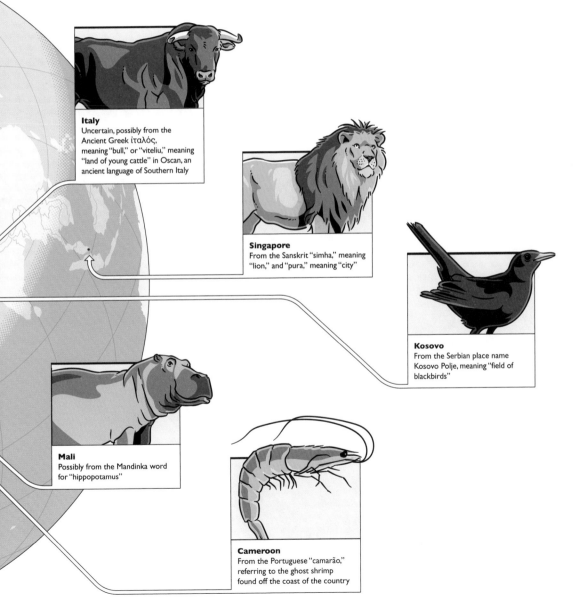

Italy
Uncertain, possibly from the Ancient Greek ἰταλός, meaning "bull," or "viteliu," meaning "land of young cattle" in Oscan, an ancient language of Southern Italy

Singapore
From the Sanskrit "simha," meaning "lion," and "pura," meaning "city"

Kosovo
From the Serbian place name Kosovo Polje, meaning "field of blackbirds"

Mali
Possibly from the Mandinka word for "hippopotamus"

Cameroon
From the Portuguese "camarão," referring to the ghost shrimp found off the coast of the country

41 The hemisphere of **water**

About 89 percent of this side of the planet is water and much of the land is under the polar ice cap.

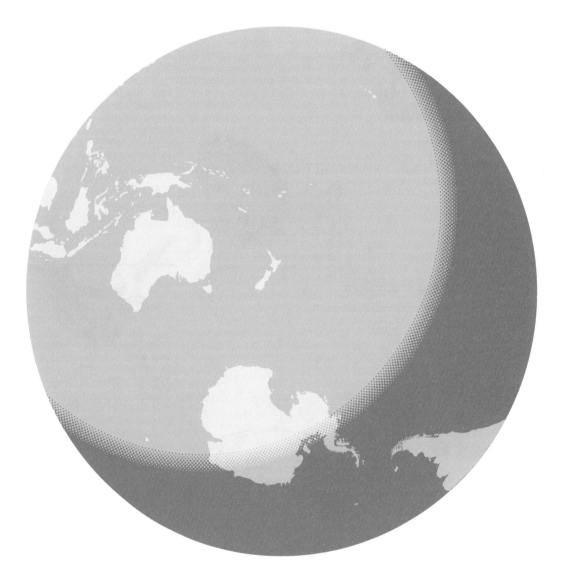

42 The hemisphere of **land**

About 80 percent of Earth's land lies on this side of the planet—nevertheless, more than half of this hemisphere is water.

43 All the **lakes** in the world

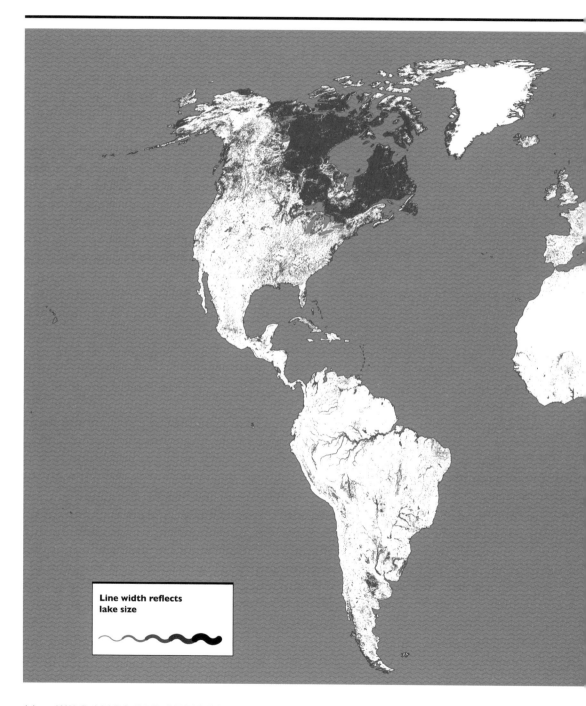

Line width reflects
lake size

There are 1.4 million lakes that are 25 acres (10 hectares) or larger. This map is drawn to emphasize the presence of lakes around the world (shown in dark blue). That's why parts of Northern Canada appear to be one enormous lake—the density of lakes there exceeds 30 percent of the landscape in certain places. The country as a whole is home to more than 60 percent of the world's lakes.

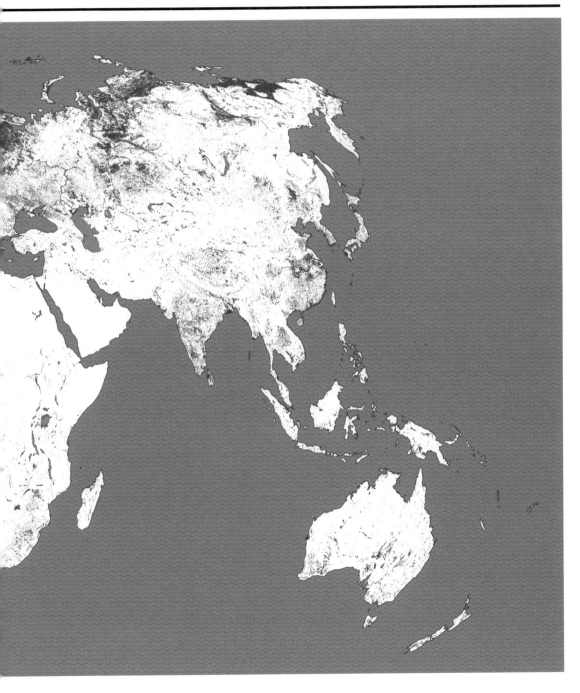

44

How the **natural world** flies its flag

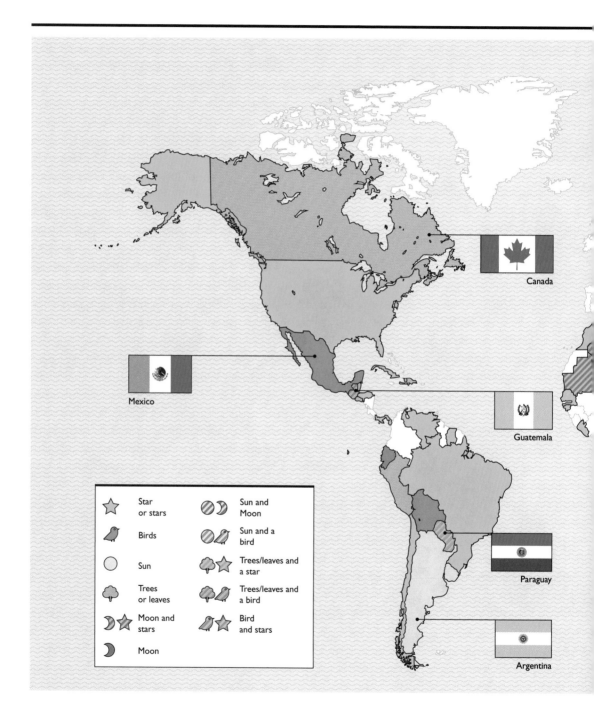

Canada

Mexico

Guatemala

Paraguay

Argentina

☆	Star or stars	◑◐	Sun and Moon
🐦	Birds	◐🐦	Sun and a bird
○	Sun	🌳☆	Trees/leaves and a star
🌳	Trees or leaves	🌳🐦	Trees/leaves and a bird
🌙☆	Moon and stars	🐦☆	Bird and stars
🌙	Moon		

Here are the most popular natural features on national flags.

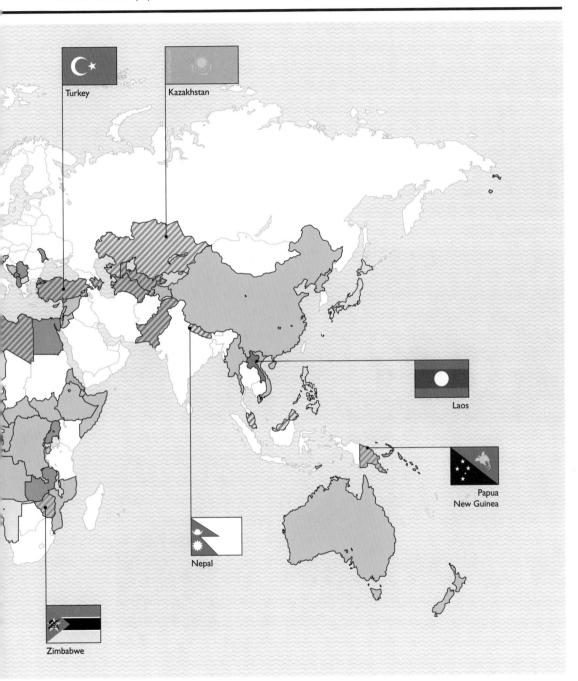

Turkey

Kazakhstan

Laos

Papua
New Guinea

Nepal

Zimbabwe

Here be **dragons**

Countries of the world with a dragon on their national flag

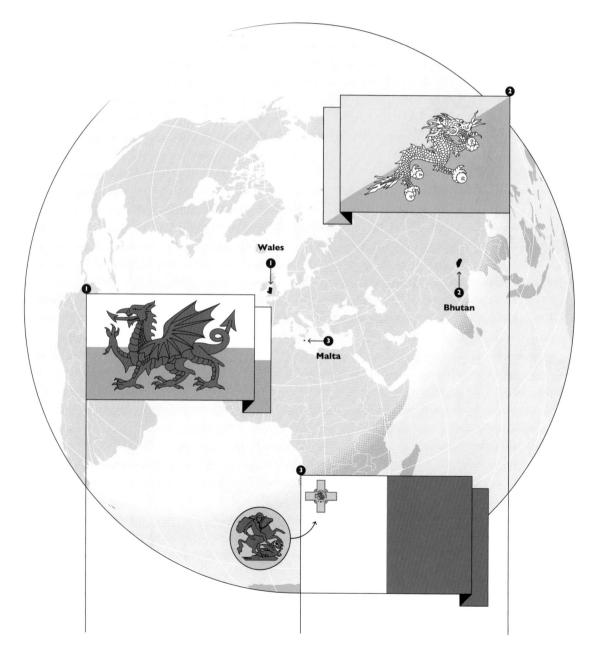

Wales ❶

Bhutan ❷

Malta ❸

The state of **land use**

The land area across the entire US used for the below purposes is closest to these states in square miles.

Mines   Golf courses   Lawns   Urban areas   Cropland

Wetlands   Wetlands lost due to development   National & State Parks   Freshwater

Impervious surfaces (pavements, roads, etc.)   Private property of fifty largest private landholders

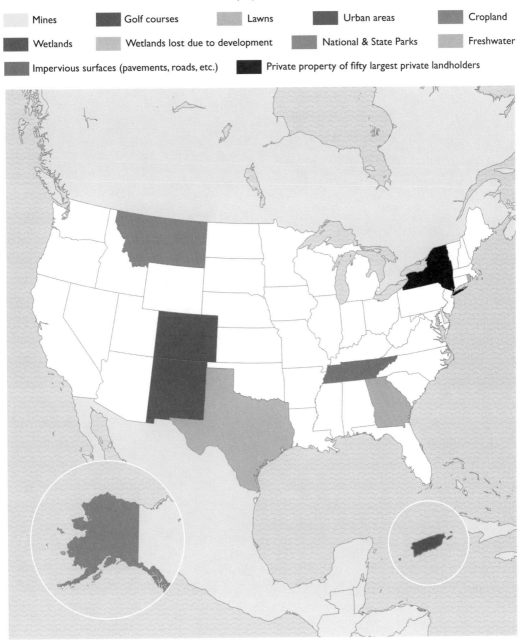

5

USING AND ABUSING NATURE

Percentage of land cultivated for
impermanent crops

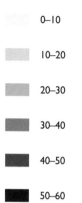

0–10

10–20

20–30

30–40

40–50

50–60

The US's Midwest may be known for its bountiful fields (think "amber waves of grain"), with nearly 400 million acres of arable land—land that is cultivated for crops such as corn, wheat, and rice that are replanted annually. No other country has more. But there's also a lot more land that *isn't* suitable for crops: Only 16.8 percent of its available land is arable, which pales in comparison to world-leading Bangladesh (59 percent), closely followed by Denmark (58.9 percent), and Ukraine (56.1 percent), the "world's breadbasket."

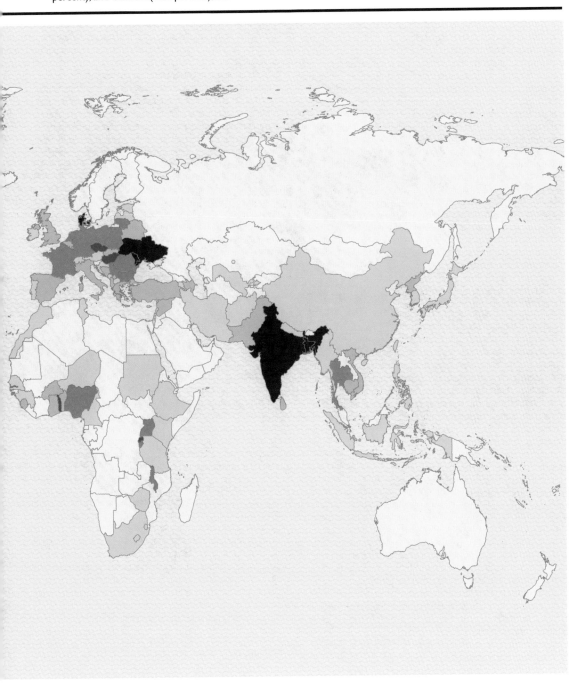

48 The **windy stretches** that could help **power the world**

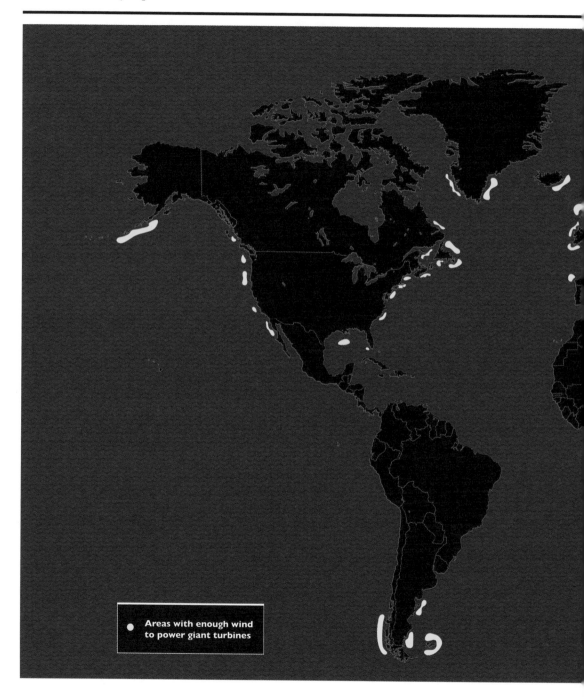

Areas with enough wind to power giant turbines

Offshore wind turbines provide about 30 gigawatts of power. Were wind to provide zero-emissions energy for all the world's needs, capacity would need to multiply by a factor of more than a thousand. That would mean erecting 2.8 million giant offshore turbines, each with a 650-foot (200 m) diameter rotor, across the windiest stretches of the world's seas.

49 The **hotspots** in Earth's crust that could help **power the world**

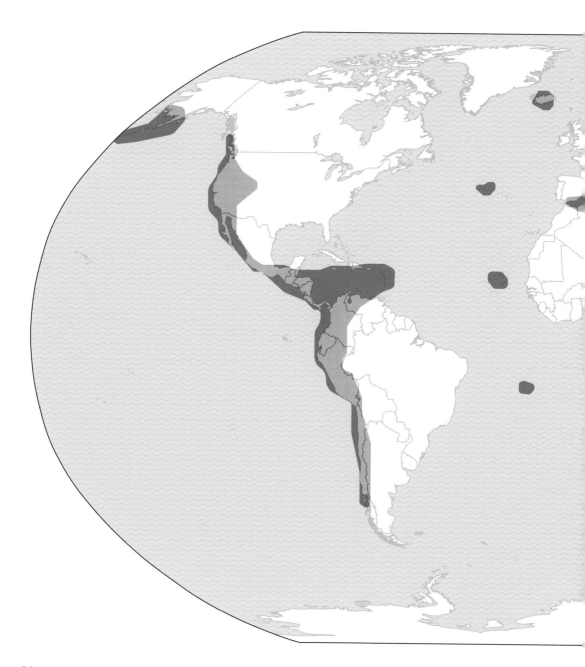

Geothermal energy might be the energy source of the future. Less than 6 miles (10 km) beneath our feet are hot rocks and water storing 50,000 times more energy than all oil and gas fields combined. And a team of scientists in Tokyo has already developed a battery to tap into the energy at Earth's crust. It is unlimited and sustainable—but converting this type of heat into electricity is still tricky.

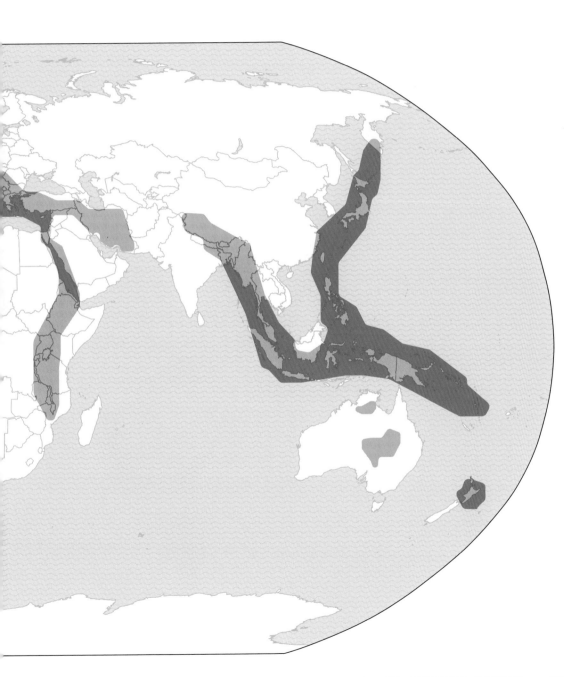

50 The **sunny places** that could help **power the world**

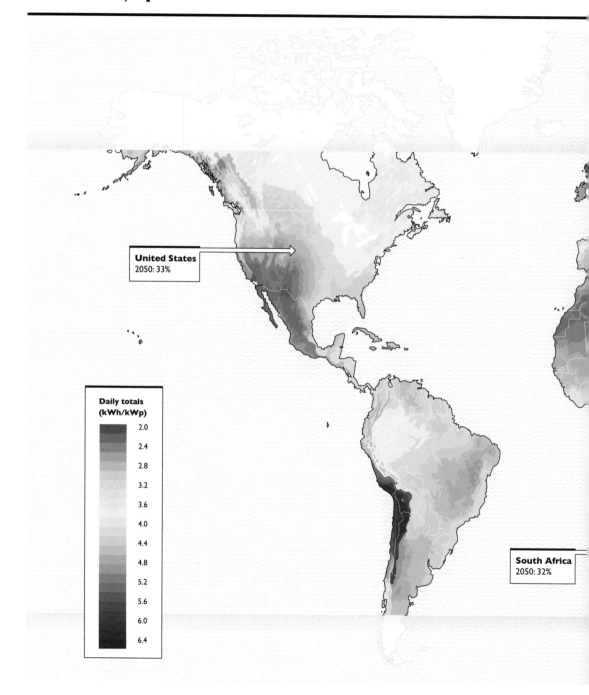

United States
2050: 33%

Daily totals (kWh/kWp)

2.0
2.4
2.8
3.2
3.6
4.0
4.4
4.8
5.2
5.6
6.0
6.4

South Africa
2050: 32%

Solar power might provide 25 percent of our energy needs by 2050—and even more than that in Australia, South Africa, and other hotspots. This map charts the long-term average of photovoltaic power potential around the world.

Japan
2050: 30%

China
2050: 23%

Australia
2050: 40%

51 National treasures: The biggest producers of gem diamonds in carats*

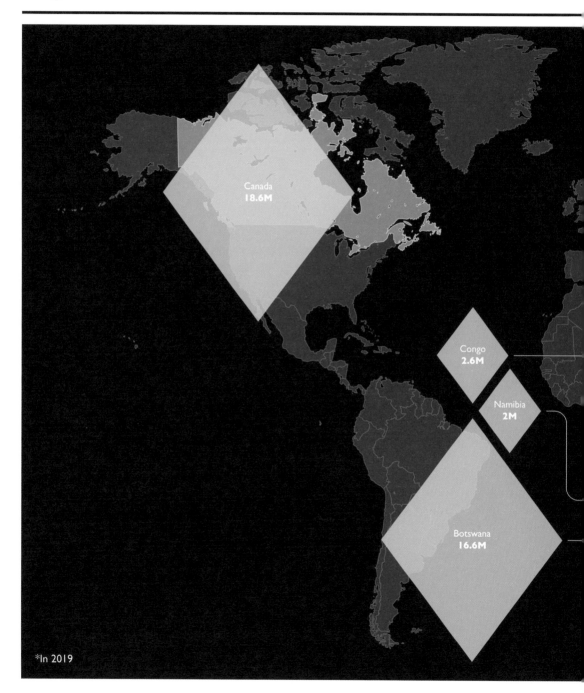

Canada
18.6M

Congo
2.6M

Namibia
2M

Botswana
16.6M

*In 2019

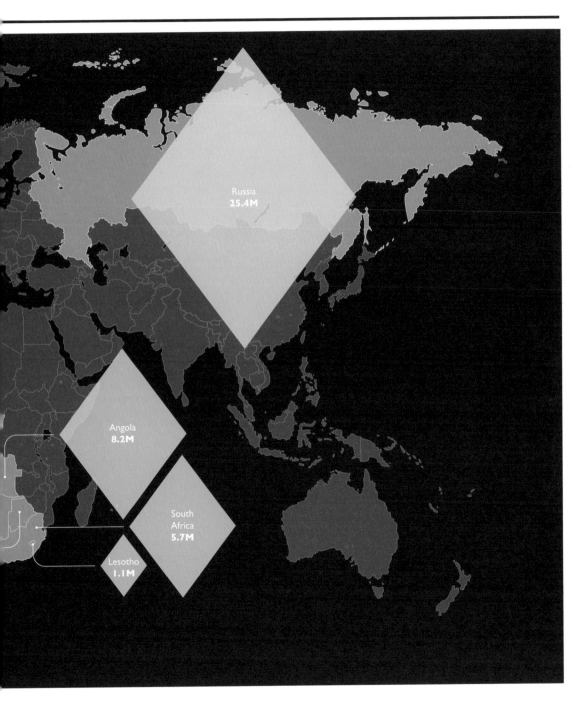

52 Who are the **gold diggers?**

Gold mined, in US tons, in 2019

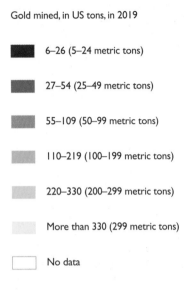

- 6–26 (5–24 metric tons)
- 27–54 (25–49 metric tons)
- 55–109 (50–99 metric tons)
- 110–219 (100–199 metric tons)
- 220–330 (200–299 metric tons)
- More than 330 (299 metric tons)
- No data

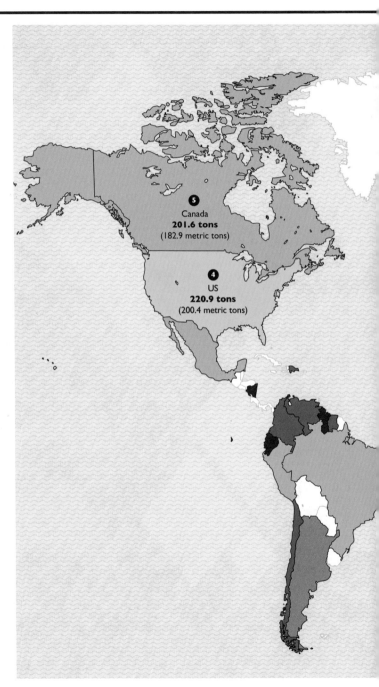

5
Canada
201.6 tons
(182.9 metric tons)

4
US
220.9 tons
(200.4 metric tons)

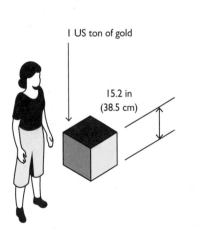

1 US ton of gold

15.2 in
(38.5 cm)

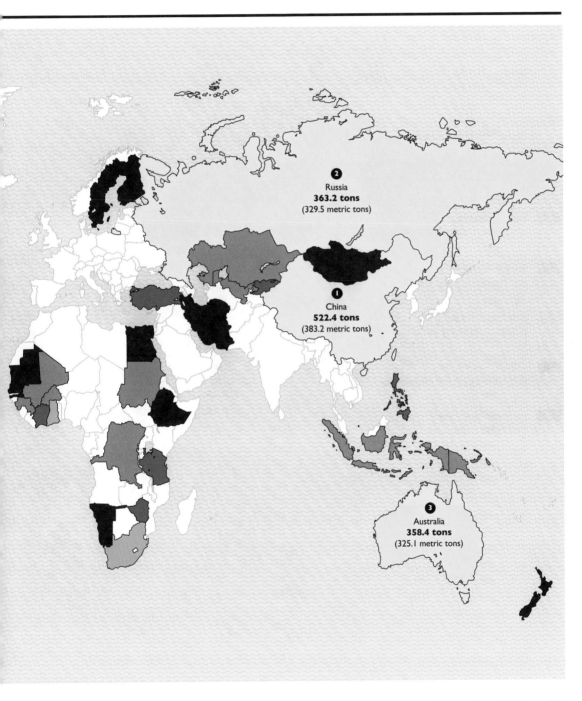

Russia
363.2 tons
(329.5 metric tons)

China
522.4 tons
(383.2 metric tons)

Australia
358.4 tons
(325.1 metric tons)

53

China grows **25 million tons of garlic** annually

Annual production, in US tons, in 2019

Less than 27,499
(25,000 metric tons)

27,500–54,999
(25,0000–49,999 metric tons)

55,000–109,999
(50,000–99,999 metric tons)

110,000–549,999
(100,000–499,999 metric tons)

550,000–1,099,999
(500,000–999,999 metric tons)

1,100,000–5,499,999
(1,000,000–4,999,999 metric tons)

5,500,000–16,499,999
(5,000,000–14,999,999 metric tons)

More than 16,499,999
(14,999,999 metric tons)

Just one country produces three times more garlic than the rest of the world combined.

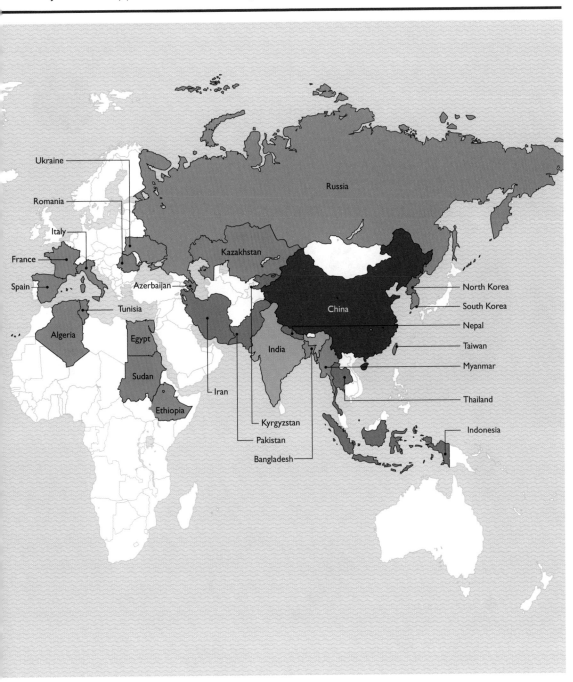

54

Who is behind the **great avocado boom?**

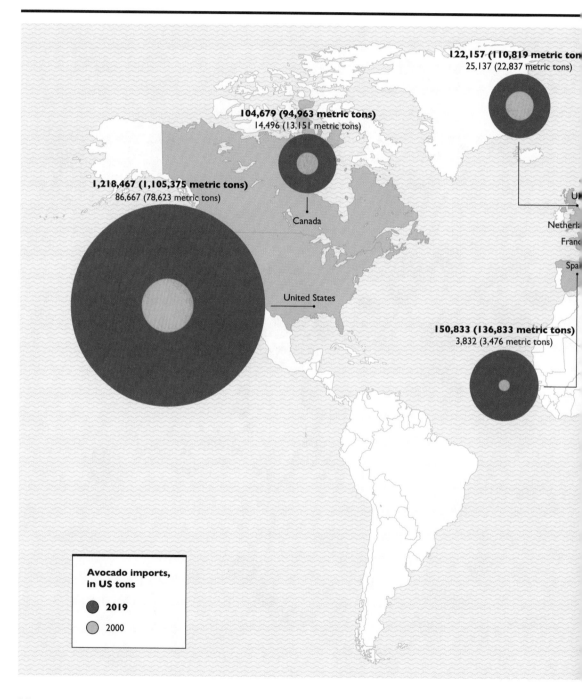

122,157 (110,819 metric tons)
25,137 (22,837 metric tons)

104,679 (94,963 metric tons)
14,496 (13,151 metric tons)

1,218,467 (1,105,375 metric tons)
86,667 (78,623 metric tons)

Canada

UK

Netherla
Franc

Spa

United States

150,833 (136,833 metric tons)
3,832 (3,476 metric tons)

**Avocado imports,
in US tons**

● 2019
● 2000

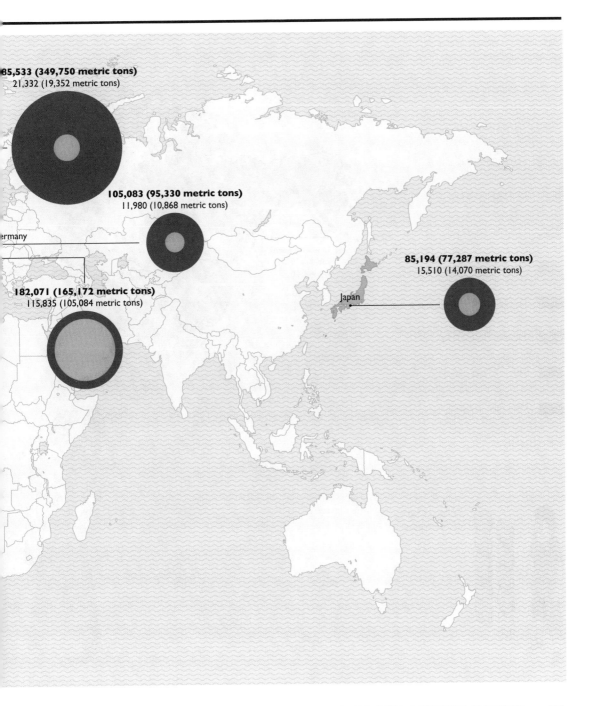

85,533 (349,750 metric tons)
21,332 (19,352 metric tons)

105,083 (95,330 metric tons)
11,980 (10,868 metric tons)

Germany

182,071 (165,172 metric tons)
115,835 (105,084 metric tons)

85,194 (77,287 metric tons)
15,510 (14,070 metric tons)

Japan

55 Who eats their **greens?**

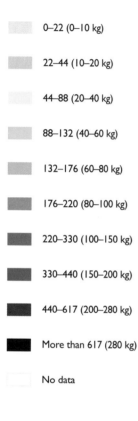

Vegetable consumption, in pounds per person, in 2019

- 0–22 (0–10 kg)
- 22–44 (10–20 kg)
- 44–88 (20–40 kg)
- 88–132 (40–60 kg)
- 132–176 (60–80 kg)
- 176–220 (80–100 kg)
- 220–330 (100–150 kg)
- 330–440 (150–200 kg)
- 440–617 (200–280 kg)
- More than 617 (280 kg)
- No data

22 lbs (10 kg) of frozen peas

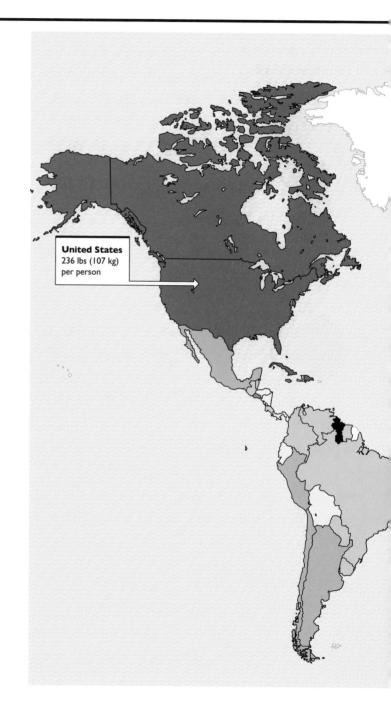

United States
236 lbs (107 kg) per person

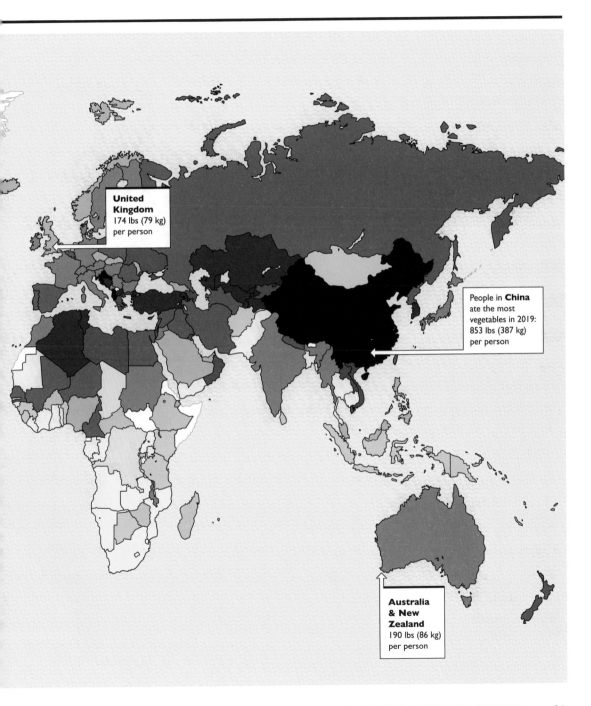

United Kingdom
174 lbs (79 kg) per person

People in **China** ate the most vegetables in 2019: 853 lbs (387 kg) per person

Australia & New Zealand
190 lbs (86 kg) per person

56 Who eats the most **fruit?**

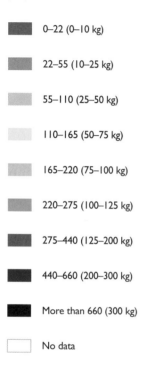

Fruit consumption, in pounds per person, in 2019

- 0–22 (0–10 kg)
- 22–55 (10–25 kg)
- 55–110 (25–50 kg)
- 110–165 (50–75 kg)
- 165–220 (75–100 kg)
- 220–275 (100–125 kg)
- 275–440 (125–200 kg)
- 440–660 (200–300 kg)
- More than 660 (300 kg)
- No data

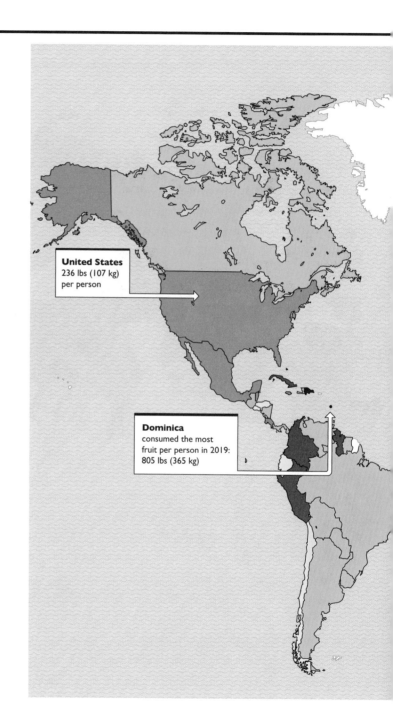

United States
236 lbs (107 kg) per person

Dominica
consumed the most fruit per person in 2019: 805 lbs (365 kg)

22 lbs (10 kg) of apples

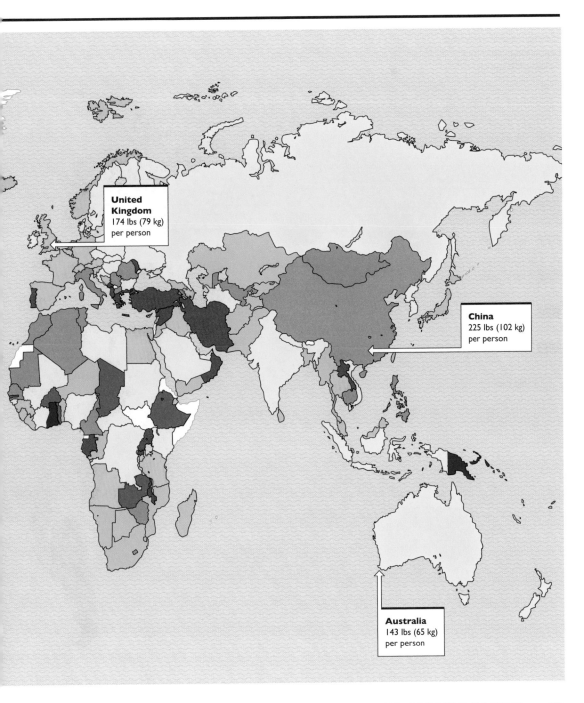

United
Kingdom
174 lbs (79 kg)
per person

China
225 lbs (102 kg)
per person

Australia
143 lbs (65 kg)
per person

57 Who eats the most **meat?**

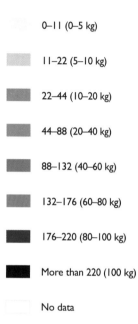

Meat consumption, in pounds per person, in 2019 (not including fish and seafood)

- 0–11 (0–5 kg)
- 11–22 (5–10 kg)
- 22–44 (10–20 kg)
- 44–88 (20–40 kg)
- 88–132 (40–60 kg)
- 132–176 (60–80 kg)
- 176–220 (80–100 kg)
- More than 220 (100 kg)
- No data

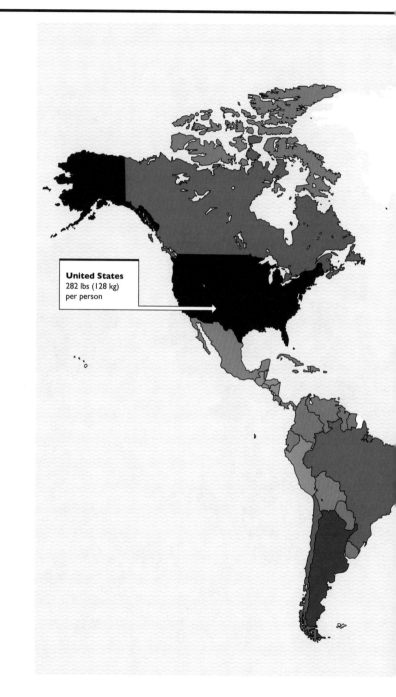

United States
282 lbs (128 kg)
per person

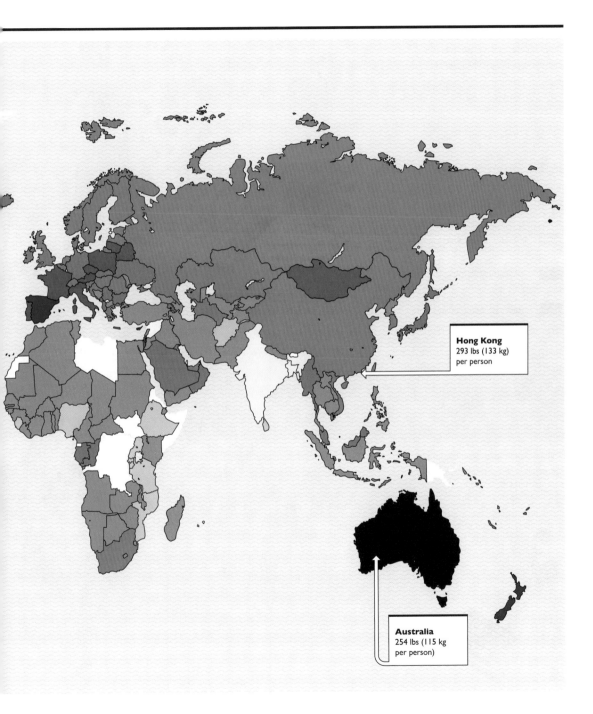

Hong Kong
293 lbs (133 kg)
per person

Australia
254 lbs (115 kg
per person)

58 The **dairy** lovers

Dairy consumption, in pounds per person, in 2019 (excluding butter)

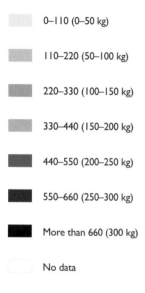

0–110 (0–50 kg)

110–220 (50–100 kg)

220–330 (100–150 kg)

330–440 (150–200 kg)

440–550 (200–250 kg)

550–660 (250–300 kg)

More than 660 (300 kg)

No data

The average Montenegrin consumed the equivalent of 6 gallons (28 L) of milk each month in 2019

28 milk cartons

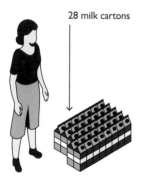

United States
514 lbs (233 kg)
per person

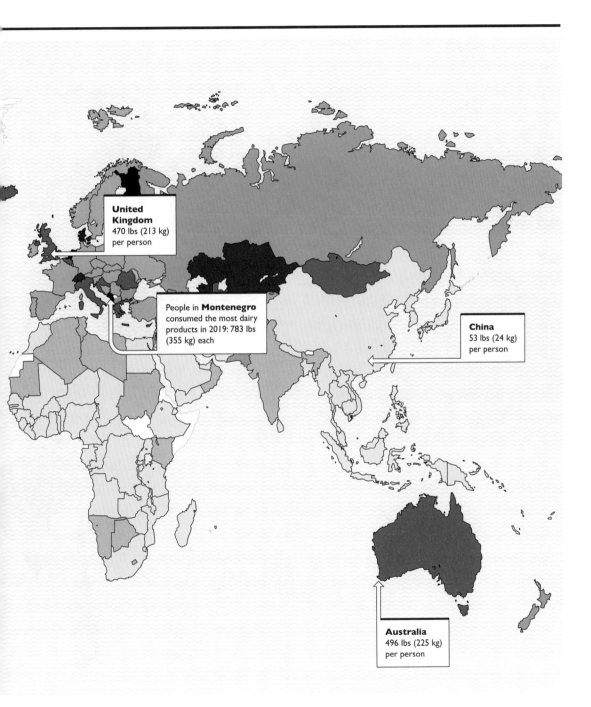

United Kingdom
470 lbs (213 kg) per person

People in **Montenegro** consumed the most dairy products in 2019: 783 lbs (355 kg) each

China
53 lbs (24 kg) per person

Australia
496 lbs (225 kg) per person

Where do all the **turkeys** live?

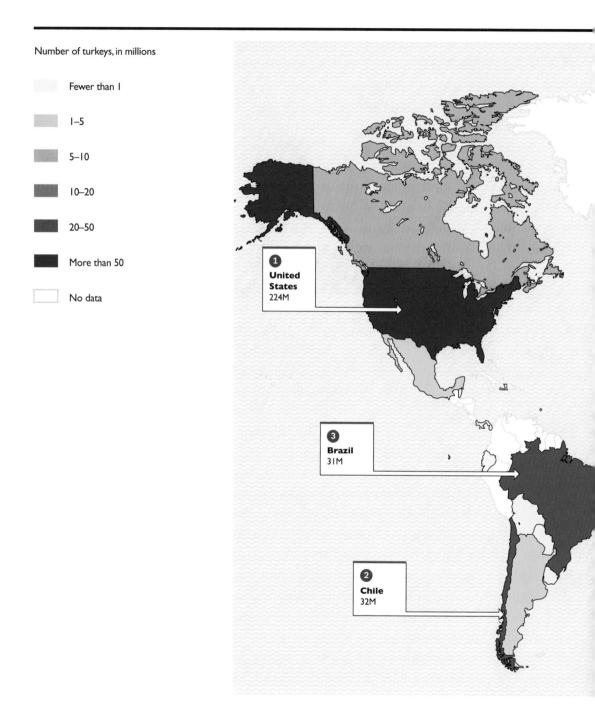

Number of turkeys, in millions

Fewer than 1

1–5

5–10

10–20

20–50

More than 50

No data

**①
United
States**
224M

**③
Brazil**
31M

**②
Chile**
32M

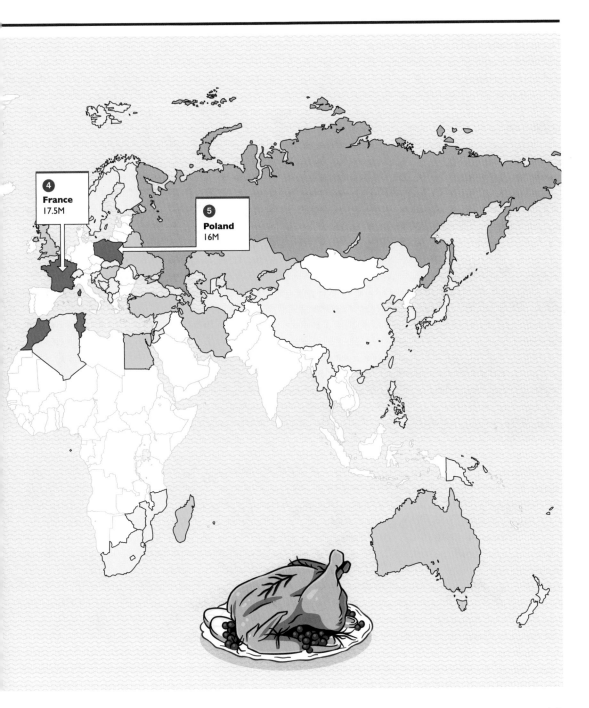

④ France
17.5M

⑤ Poland
16M

60 Who has **killed whales** since the 1985 ban?

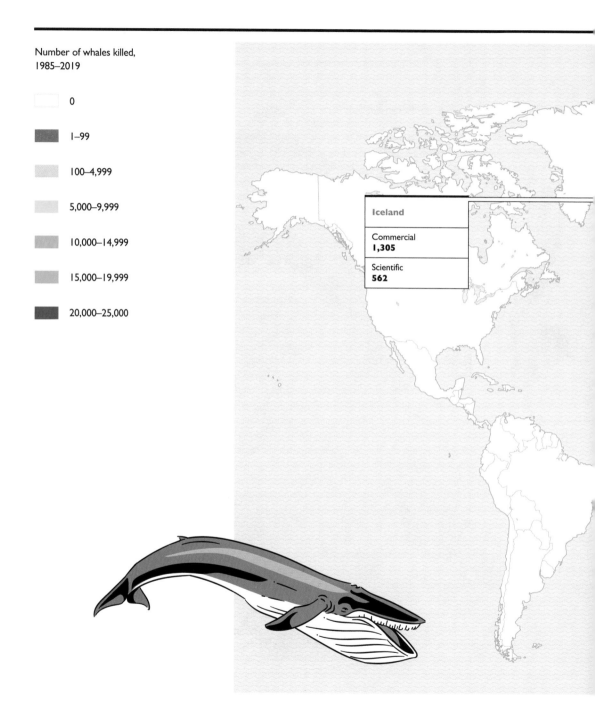

Number of whales killed,
1985–2019

0

1–99

100–4,999

5,000–9,999

10,000–14,999

15,000–19,999

20,000–25,000

Iceland

Commercial
1,305

Scientific
562

These countries have hunted whales commercially or for science since the International Whaling Commission declared a worldwide moratorium in 1985.

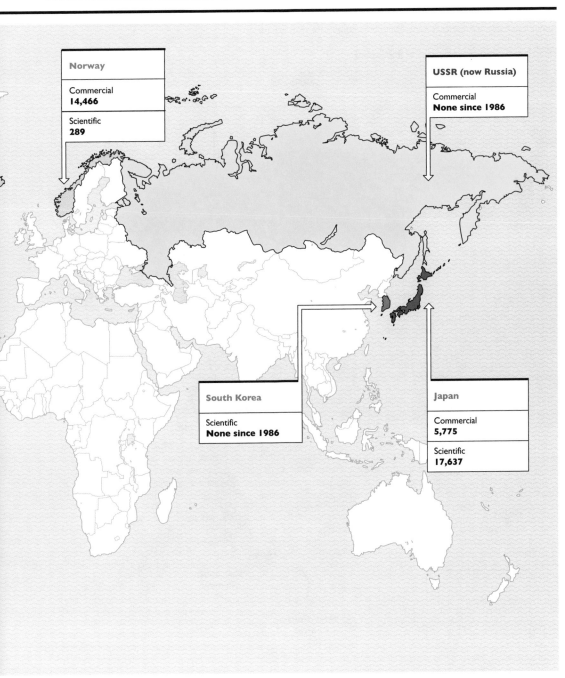

Norway

Commercial
14,466

Scientific
289

USSR (now Russia)

Commercial
None since 1986

South Korea

Scientific
None since 1986

Japan

Commercial
5,775

Scientific
17,637

Where do **America's hunters** live?

About 4 percent of Americans hunt—but some hunt more than others.

Hunting participation (percentage of population)

1–2   3–4   5–6   7–8

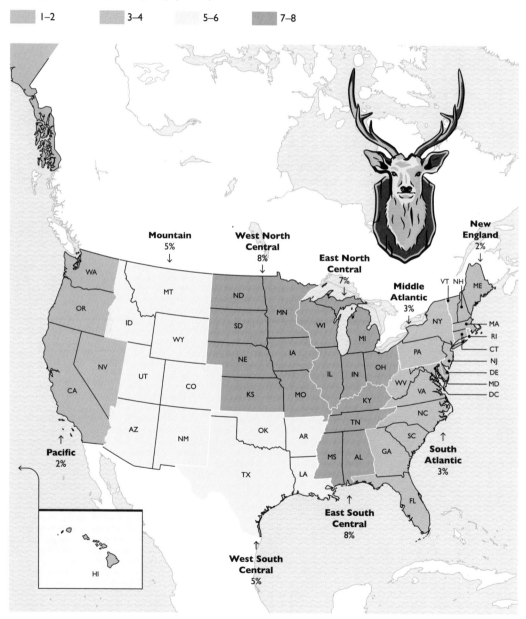

Mountain
5%
↓

West North
Central
8%
↓

East North
Central
7%
↓

Middle
Atlantic
3%
↓

New
England
2%
↓

Pacific
2%
↑

South
Atlantic
3%
↑

East South
Central
8%
↑

West South
Central
5%
↑

62 Where **rhinos** have been poached

There are just 27,000 rhinos left in the wild, down from half a million a century ago.

Total number of rhinos poached, 1990–2017

1–99   100–199   200–999   1,000–4,999   5,000 or more

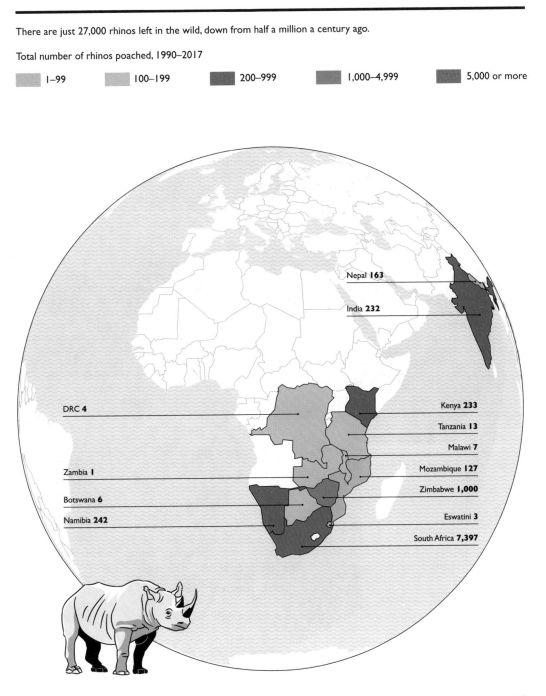

Nepal **163**

India **232**

DRC **4**

Kenya **233**

Tanzania **13**

Malawi **7**

Mozambique **127**

Zambia **1**

Zimbabwe **1,000**

Botswana **6**

Namibia **242**

Eswatini **3**

South Africa **7,397**

63 Where American ships **killed whales** in the 19th century

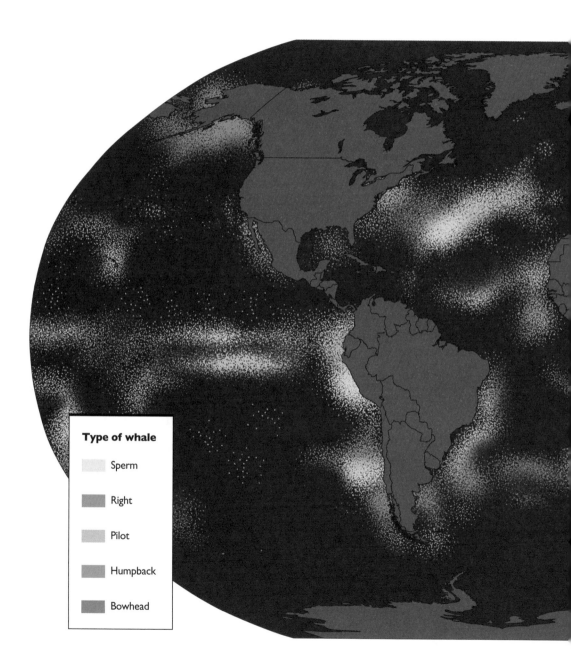

Type of whale

- Sperm
- Right
- Pilot
- Humpback
- Bowhead

Whaling ships accounted for the vast majority of the global fleet, and whaling at its height was America's fifth largest industry. Whale oil of various kinds provided high-quality lighting and lubrication as well as other valuable products.

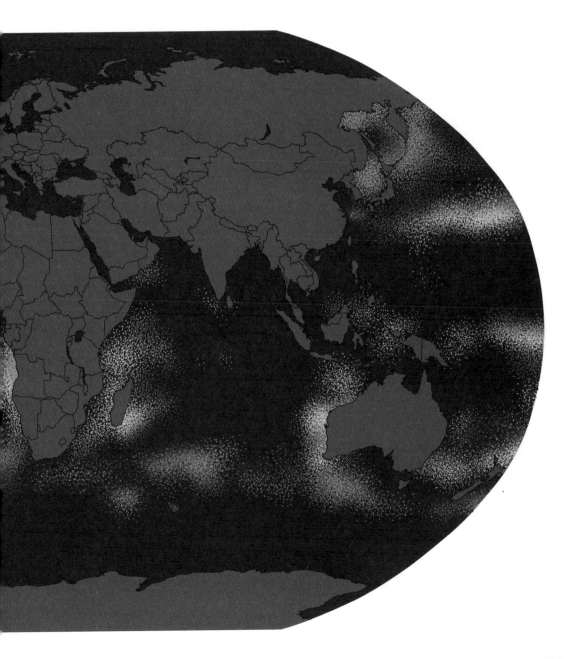

64 Who are the world's **cat people?**

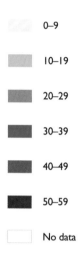

Percentage of those polled in
each country who said they
owned a cat

0–9

10–19

20–29

30–39

40–49

50–59

No data

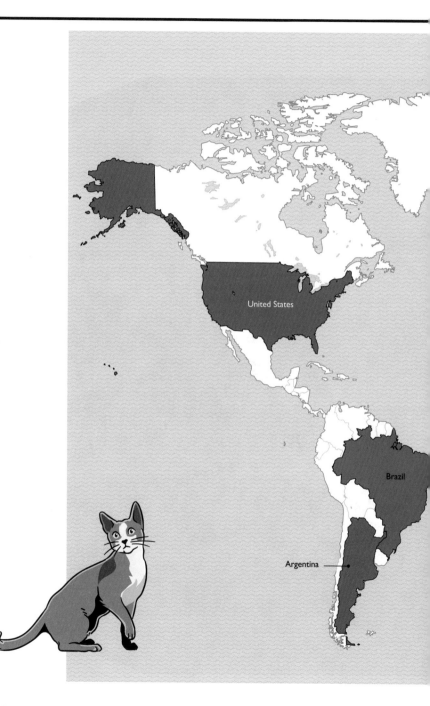

United States

Brazil

Argentina

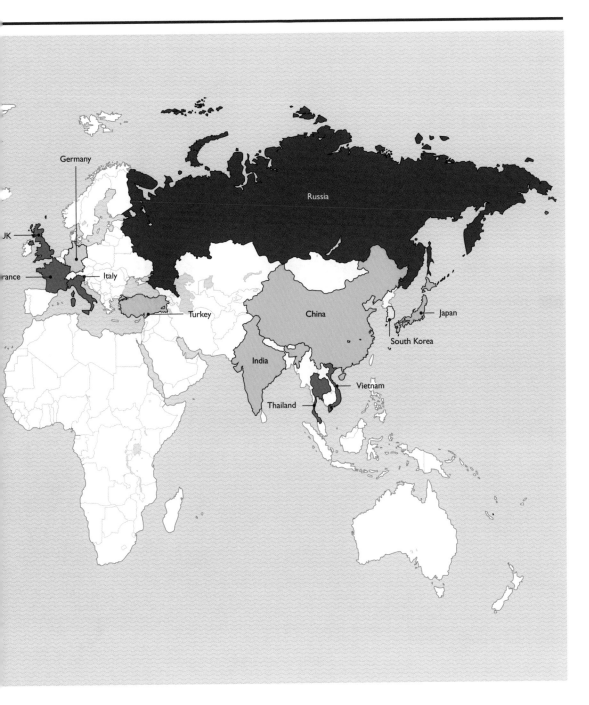

65 Which countries are responsible for the most types and breeds of **dog?**

Number of breeds and types
of dog

0

1–10

11–20

21–30

31–40

41–50

More than 50

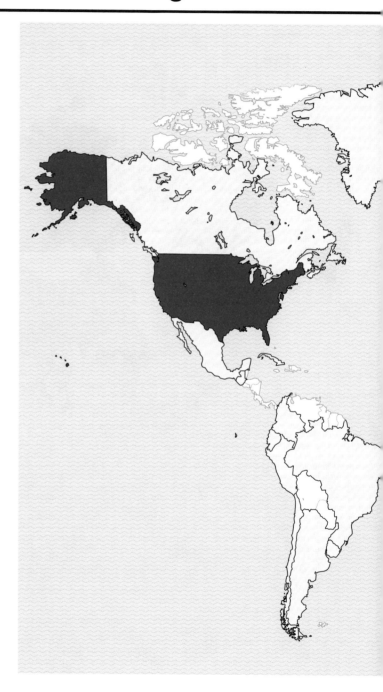

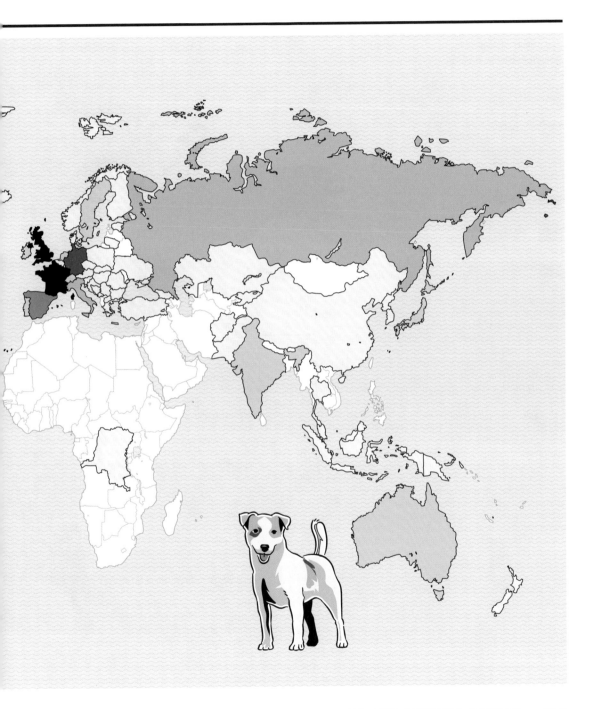

66 Who has kept **dolphins** in captivity?

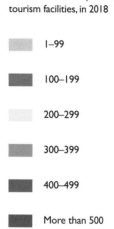

The number of dolphins in tourism facilities, in 2018

- 1–99
- 100–199
- 200–299
- 300–399
- 400–499
- More than 500

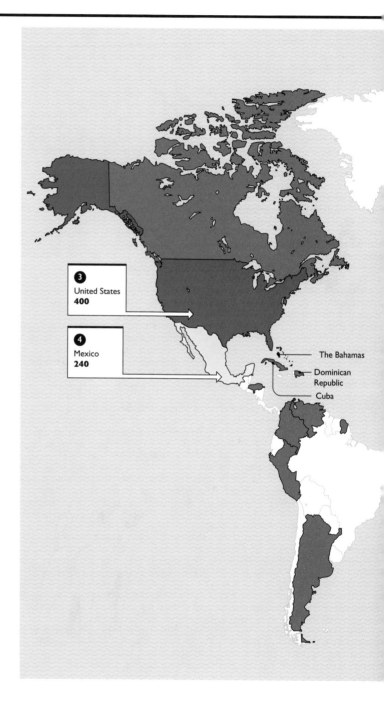

3 United States **400**

4 Mexico **240**

The Bahamas

Dominican Republic

Cuba

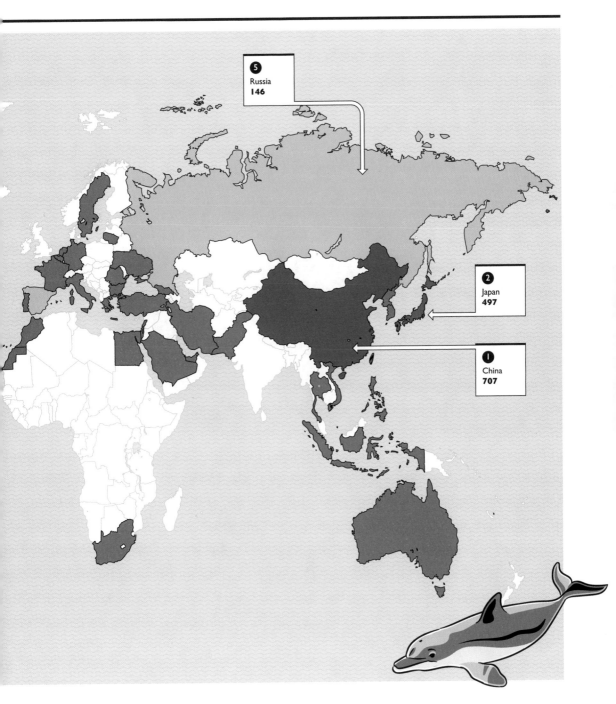

⑤ Russia
146

② Japan
497

① China
707

6

EXTREME EARTH

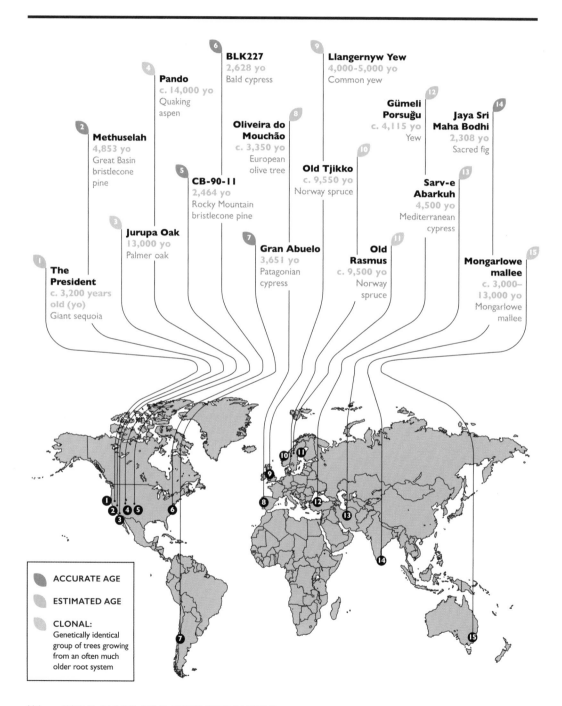

6 BLK227
2,628 yo
Bald cypress

9 Llangernyw Yew
4,000-5,000 yo
Common yew

4 Pando
c. 14,000 yo
Quaking
aspen

**12 Gümeli
Porsuğu**
c. 4,115 yo
Yew

**14 Jaya Sri
Maha Bodhi**
2,308 yo
Sacred fig

**8 Oliveira do
Mouchão**
c. 3,350 yo
European
olive tree

2 Methuselah
4,853 yo
Great Basin
bristlecone
pine

10 Old Tjikko
c. 9,550 yo
Norway spruce

5 CB-90-11
2,464 yo
Rocky Mountain
bristlecone
pine

**13 Sarv-e
Abarkuh**
4,500 yo
Mediterranean
cypress

3 Jurupa Oak
13,000 yo
Palmer oak

7 Gran Abuelo
3,651 yo
Patagonian
cypress

**11 Old
Rasmus**
c. 9,500 yo
Norway
spruce

**15 Mongarlowe
mallee**
c. 3,000–
13,000 yo
Mongarlowe
mallee

**1 The
President**
c. 3,200 years
old (yo)
Giant sequoia

ACCURATE AGE

ESTIMATED AGE

CLONAL:
Genetically identical
group of trees growing
from an often much
older root system

68 The **highs** and **lows** of North America

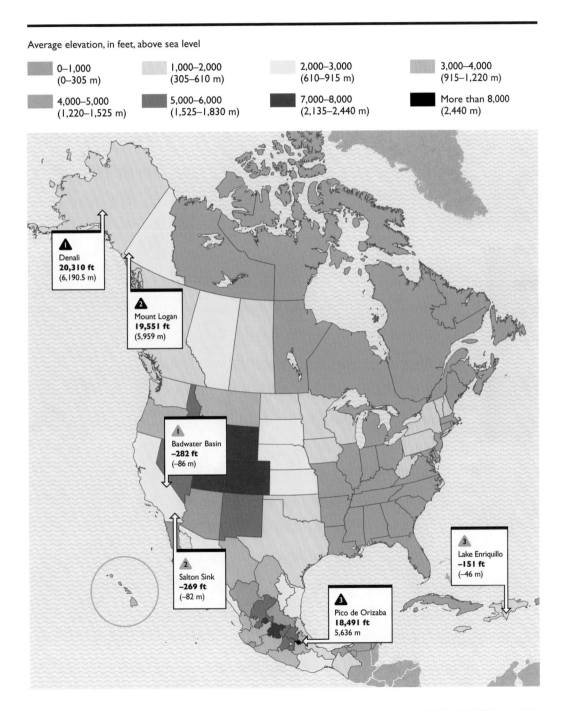

Average elevation, in feet, above sea level

- 0–1,000 (0–305 m)
- 1,000–2,000 (305–610 m)
- 2,000–3,000 (610–915 m)
- 3,000–4,000 (915–1,220 m)
- 4,000–5,000 (1,220–1,525 m)
- 5,000–6,000 (1,525–1,830 m)
- 7,000–8,000 (2,135–2,440 m)
- More than 8,000 (2,440 m)

△1 Denali
20,310 ft
(6,190.5 m)

△2 Mount Logan
19,551 ft
(5,959 m)

▽1 Badwater Basin
−282 ft
(−86 m)

▽2 Salton Sink
−269 ft
(−82 m)

▽3 Lake Enriquillo
−151 ft
(−46 m)

△3 Pico de Orizaba
18,491 ft
5,636 m

69 **Warning!** Dangerous animals at work

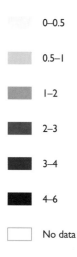

The number of people who die by venomous bite or sting (per 100,000 humans)

- 0–0.5
- 0.5–1
- 1–2
- 2–3
- 3–4
- 4–6
- No data

Where are you most likely to be killed by a venomous animal? Somalia and India stand out here. As of 2019, on average, for every 100,000 Somalians, 5.62 are killed annually by venom, making it the world's most dangerous country for deadly bites or stings. In India, the rate is 4.56 people, which is, although still very high, trending downward; in 1990, it ranked the highest in the world at 9.09 deaths per 100,000 people.

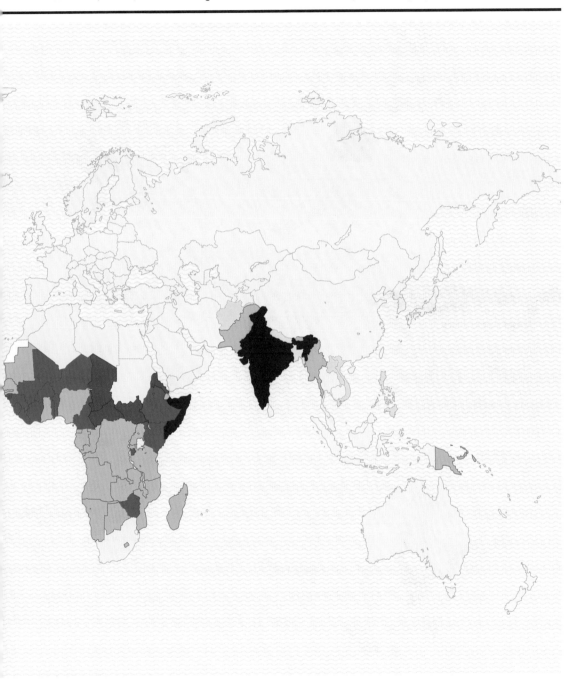

Watch your step—these creatures are the **tiniest** of their kind

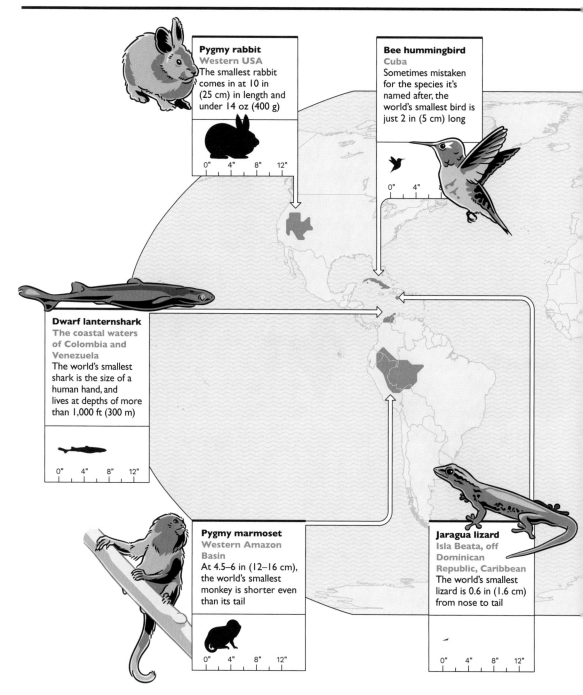

Pygmy rabbit
Western USA
The smallest rabbit comes in at 10 in (25 cm) in length and under 14 oz (400 g)

0" 4" 8" 12"

Bee hummingbird
Cuba
Sometimes mistaken for the species it's named after, the world's smallest bird is just 2 in (5 cm) long

0" 4" 8"

Dwarf lanternshark
The coastal waters of Colombia and Venezuela
The world's smallest shark is the size of a human hand, and lives at depths of more than 1,000 ft (300 m)

0" 4" 8" 12"

Pygmy marmoset
Western Amazon Basin
At 4.5–6 in (12–16 cm), the world's smallest monkey is shorter even than its tail

0" 4" 8" 12"

Jaragua lizard
Isla Beata, off Dominican Republic, Caribbean
The world's smallest lizard is 0.6 in (1.6 cm) from nose to tail

0" 4" 8" 12"

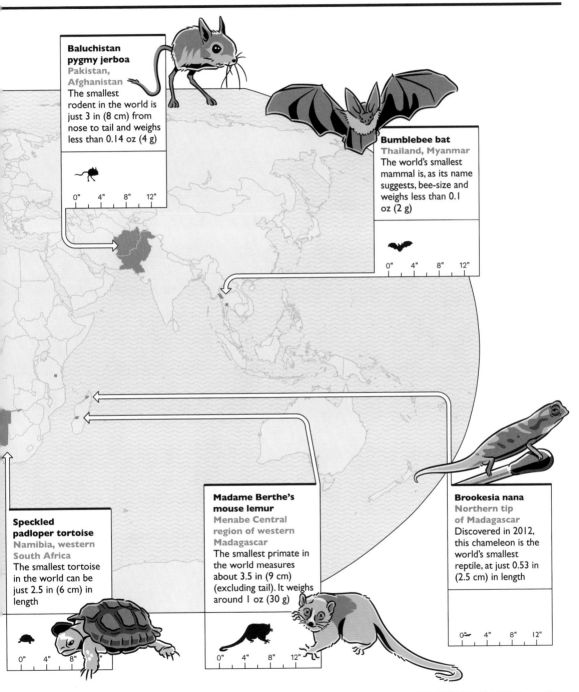

Baluchistan pygmy jerboa
Pakistan, Afghanistan
The smallest rodent in the world is just 3 in (8 cm) from nose to tail and weighs less than 0.14 oz (4 g)

0" 4" 8" 12"

Bumblebee bat
Thailand, Myanmar
The world's smallest mammal is, as its name suggests, bee-size and weighs less than 0.1 oz (2 g)

0" 4" 8" 12"

Speckled padloper tortoise
Namibia, western South Africa
The smallest tortoise in the world can be just 2.5 in (6 cm) in length

0" 4" 8" 12"

Madame Berthe's mouse lemur
Menabe Central region of western Madagascar
The smallest primate in the world measures about 3.5 in (9 cm) (excluding tail). It weighs around 1 oz (30 g)

0" 4" 8" 12"

Brookesia nana
Northern tip of Madagascar
Discovered in 2012, this chameleon is the world's smallest reptile, at just 0.53 in (2.5 cm) in length

0" 4" 8" 12"

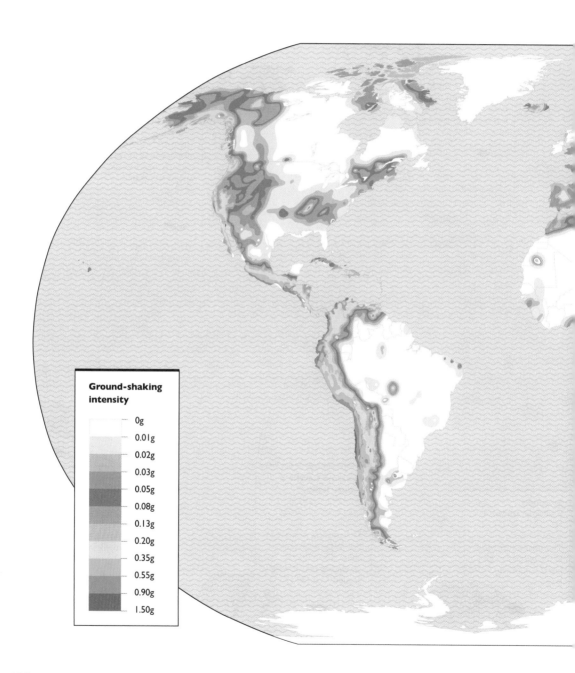

71 How far are you from an **earthquake zone?**

Ground-shaking intensity

- 0g
- 0.01g
- 0.02g
- 0.03g
- 0.05g
- 0.08g
- 0.13g
- 0.20g
- 0.35g
- 0.55g
- 0.90g
- 1.50g

This map shows the ground-shaking intensity caused by earthquakes that we might reasonably expect to occur once every 475 years. The shaking is measured in "g," the acceleration due to the Earth's gravity; for instance, 0.02g (i.e., an accleration equal to 2% of the gravitational acceleration) indicates acceleration produced by low-strength quakes, while 0.9g, or 90%g, indicates the acceleration produced by a violent quake.

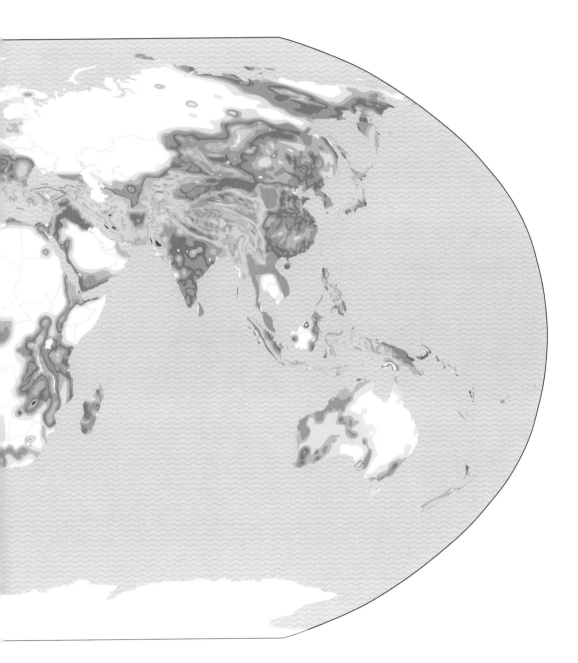

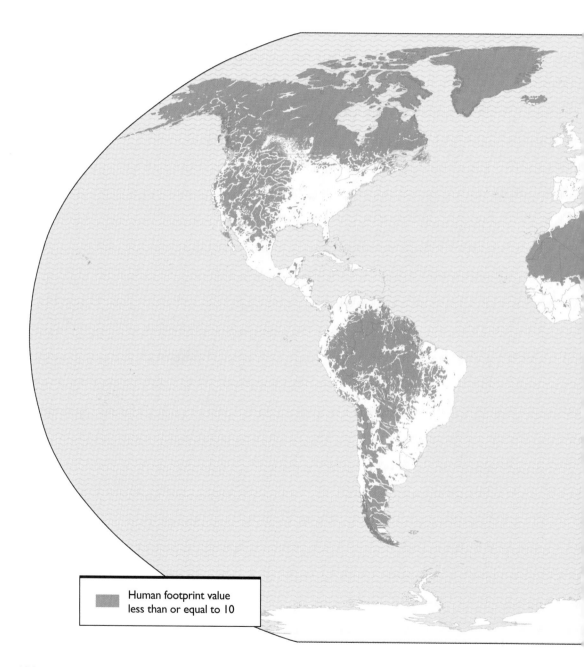

Human footprint value
less than or equal to 10

Another way of identifying where wild land has been left intact is to find the places where there aren't (many) humans around. When a team of researchers studied the worldwide human footprint, they rated each area's human impact on a scale of 1 (least) to 100 (most) and found that 83 percent of Earth's land surface is influenced by human activity. This can have dire consequences for wildlife; human-caused habitat fragmentation and loss is possibly the leading factor in loss of biodiversity. This map represents the remaining lands with minimal human impact (a rating of 10 or below).

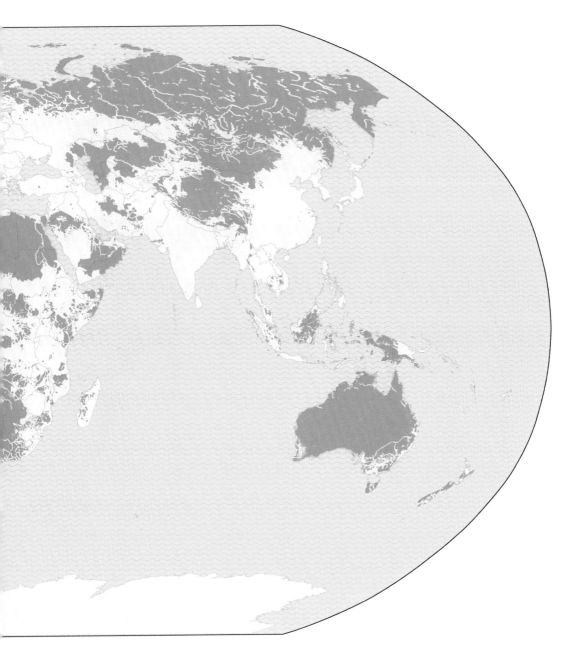

73 Where are people at **most risk** from **natural disasters?**

Current risk of exposure to earthquakes, hurricanes, floods, droughts, and sea-level rise. The countries with greatest risk are ranked 1–10.

No data

Very low

Low

Medium

High

Very high

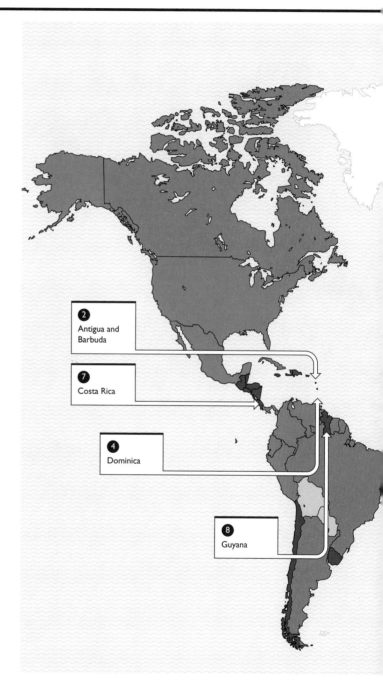

2 Antigua and Barbuda

7 Costa Rica

4 Dominica

8 Guyana

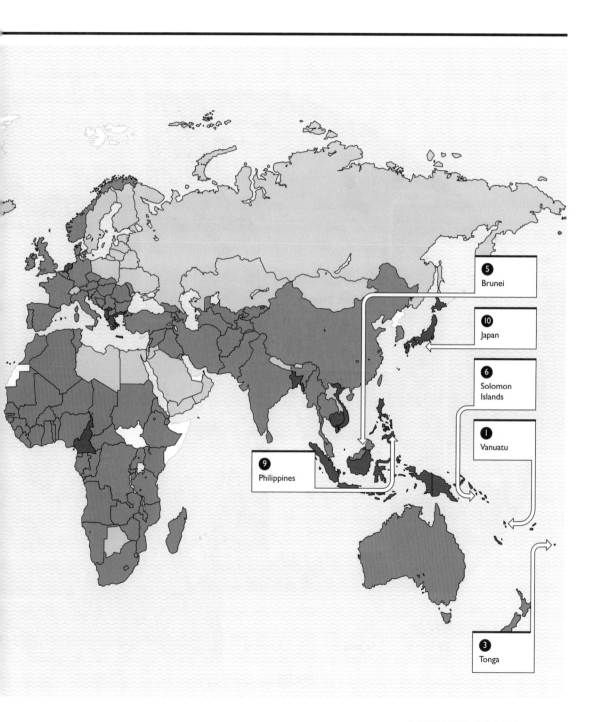

5 Brunei

10 Japan

6 Solomon Islands

1 Vanuatu

9 Philippines

3 Tonga

Ocean deep, **mountain** high

Tonga Trench
The fastest tectonic plate movement on Earth occurs in the 35,486-ft (10,816 m) deep Tonga Trench—9.5 in (24 cm) per year in certain places

Aleutian Trench
In 1946, a tsunami originating from along the Aleutian Trench struck Hawaii, killing 173 people

Andes
Stretching more than 4,000 mi (7,000 km), the Andes is the longest continental mountain range in the world, and the highest outside Asia

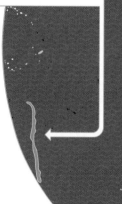

Elevation		Depth	
16,400 ft (5,000 m)	**29,200 ft** (8,900 m)	**36,000 ft** (11,000 m)	**21,000 ft** (6,500 m)

Between Earth's most elevated point—the summit of Mount Everest, 29,032 feet (8,848 m)—and its deepest—the Mariana Trench, 36,200 feet (11,034 m)—is a vertical distance of nearly 12.5 miles (20 km).

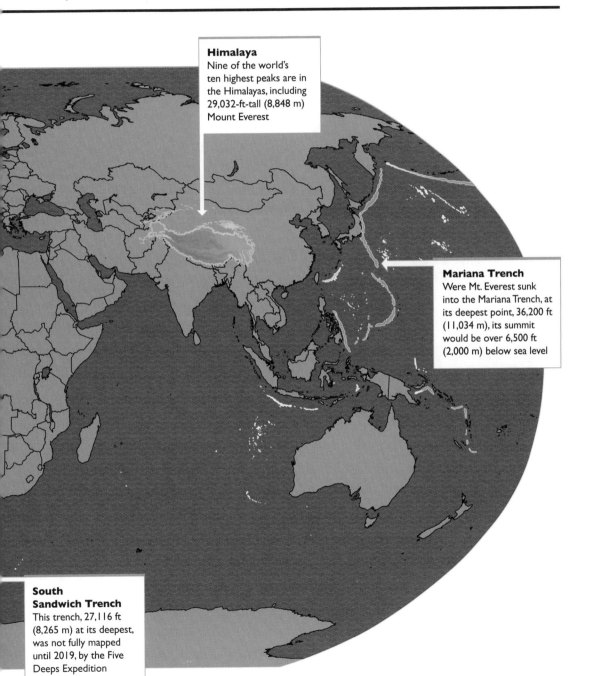

Himalaya
Nine of the world's ten highest peaks are in the Himalayas, including 29,032-ft-tall (8,848 m) Mount Everest

Mariana Trench
Were Mt. Everest sunk into the Mariana Trench, at its deepest point, 36,200 ft (11,034 m), its summit would be over 6,500 ft (2,000 m) below sea level

South Sandwich Trench
This trench, 27,116 ft (8,265 m) at its deepest, was not fully mapped until 2019, by the Five Deeps Expedition

The **largest iceberg** reliably recorded was bigger than Corsica and Mallorca

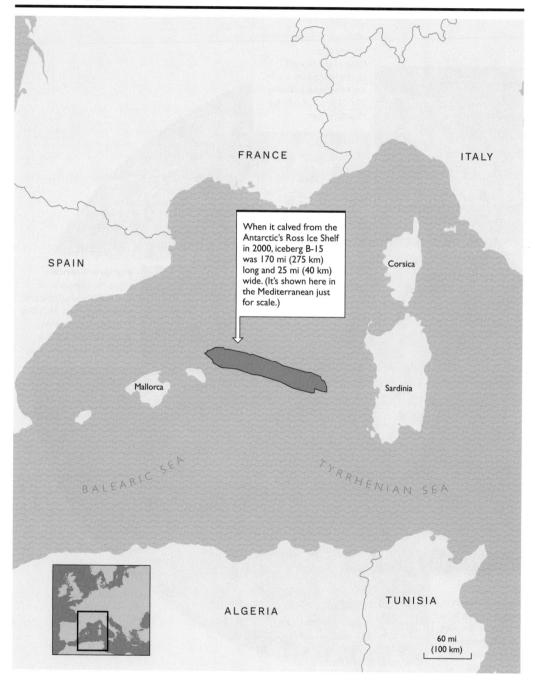

FRANCE

ITALY

SPAIN

When it calved from the Antarctic's Ross Ice Shelf in 2000, iceberg B-15 was 170 mi (275 km) long and 25 mi (40 km) wide. (It's shown here in the Mediterranean just for scale.)

Corsica

Mallorca

Sardinia

BALEARIC SEA

TYRRHENIAN SEA

ALGERIA

TUNISIA

60 mi
(100 km)

How people have **died** in **America's national parks**

Legend:
- Visitors, in 2019
- Drowning
- Motor vehicle crash
- Falling
- Transportation
- Environmental
- Poisoning
- Wildlife/animal
- Other
- Medical/natural death
- Homicide
- Undetermined
- Total deaths (2010–2020)

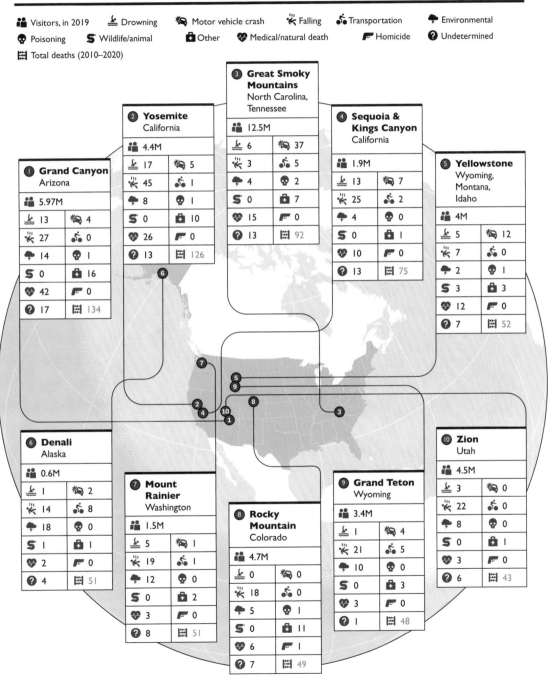

③ Great Smoky Mountains
North Carolina, Tennessee

Visitors: 12.5M

Drowning	6	Motor vehicle crash	37
Falling	3	Transportation	5
Environmental	4	Poisoning	2
Wildlife/animal	0	Other	7
Medical/natural death	15	Homicide	0
Undetermined	13	Total deaths	92

② Yosemite
California

Visitors: 4.4M

Drowning	17	Motor vehicle crash	5
Falling	45	Transportation	1
Environmental	8	Poisoning	1
Wildlife/animal	0	Other	10
Medical/natural death	26	Homicide	0
Undetermined	13	Total deaths	126

④ Sequoia & Kings Canyon
California

Visitors: 1.9M

Drowning	13	Motor vehicle crash	7
Falling	25	Transportation	2
Environmental	4	Poisoning	0
Wildlife/animal	0	Other	1
Medical/natural death	10	Homicide	0
Undetermined	13	Total deaths	75

① Grand Canyon
Arizona

Visitors: 5.97M

Drowning	13	Motor vehicle crash	4
Falling	27	Transportation	0
Environmental	14	Poisoning	1
Wildlife/animal	0	Other	16
Medical/natural death	42	Homicide	0
Undetermined	17	Total deaths	134

⑤ Yellowstone
Wyoming, Montana, Idaho

Visitors: 4M

Drowning	5	Motor vehicle crash	12
Falling	7	Transportation	0
Environmental	2	Poisoning	1
Wildlife/animal	3	Other	3
Medical/natural death	12	Homicide	0
Undetermined	7	Total deaths	52

⑥ Denali
Alaska

Visitors: 0.6M

Drowning	1	Motor vehicle crash	2
Falling	14	Transportation	8
Environmental	18	Poisoning	0
Wildlife/animal	1	Other	1
Medical/natural death	2	Homicide	0
Undetermined	4	Total deaths	51

⑦ Mount Rainier
Washington

Visitors: 1.5M

Drowning	5	Motor vehicle crash	1
Falling	19	Transportation	1
Environmental	12	Poisoning	0
Wildlife/animal	0	Other	2
Medical/natural death	3	Homicide	0
Undetermined	8	Total deaths	51

⑧ Rocky Mountain
Colorado

Visitors: 4.7M

Drowning	0	Motor vehicle crash	0
Falling	18	Transportation	0
Environmental	5	Poisoning	1
Wildlife/animal	0	Other	11
Medical/natural death	6	Homicide	1
Undetermined	7	Total deaths	49

⑨ Grand Teton
Wyoming

Visitors: 3.4M

Drowning	1	Motor vehicle crash	4
Falling	21	Transportation	5
Environmental	10	Poisoning	0
Wildlife/animal	0	Other	3
Medical/natural death	3	Homicide	0
Undetermined	1	Total deaths	48

⑩ Zion
Utah

Visitors: 4.5M

Drowning	3	Motor vehicle crash	0
Falling	22	Transportation	0
Environmental	8	Poisoning	0
Wildlife/animal	0	Other	1
Medical/natural death	3	Homicide	0
Undetermined	6	Total deaths	43

Of the 66 tornadoes widely accepted to have been rated F5/EF5 from first recording through to June 2022, 59 have occurred in the US, especially the middle of the country, where the perfect ingredients for tornadoes (cold, dry air meeting warm, humid air) can be found every spring.

○ F5/EF5 tornadoes

Areas with highest frequency of tornadoes

Tornadoes are a global phenomenon, with high-risk areas on every continent besides Antarctica. (And they usually do rotate in opposite directions in the northern and southern hemispheres—counterclockwise and clockwise, respectively.) But a closer look tells a startling story: The US is an extreme outlier, enduring by far the most tornadoes each year (roughly 1,200 annually; Canada comes second, with about 100 tornadoes a year) as well as the majority of the strongest (F5/EF5) tornadoes.

7

THE PLANET
IN PERIL

78 How many acres **burn** in the US each year?

This map compares how much land burned in an average year from 1984–2001 to how much land burned in an average year from 2002–2018; 37 out of 50 states saw more destruction in the latter span. The key indicates how many more (or fewer) acres burned per square mile. One square mile contains 640 acres.

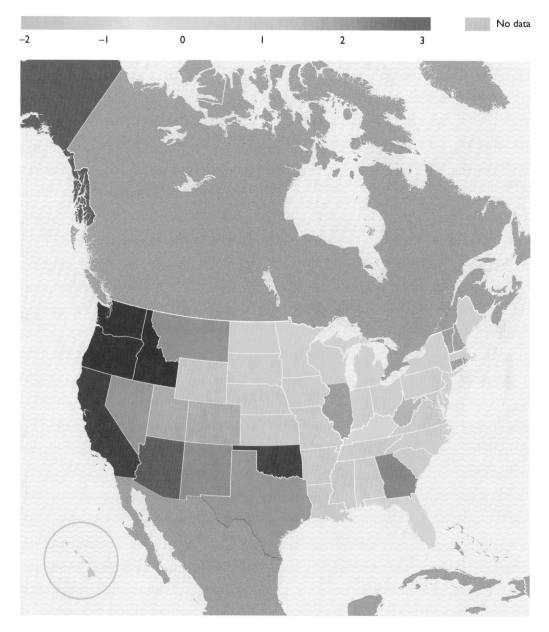

No data

−2   −1   0   1   2   3

The **Chernobyl disaster** led to a nature reserve **six times** the size of New York City

In 1986, after the catastrophic explosion of the nuclear power plant at Chernobyl, Ukraine, around 200,000 people were evacuated from contaminated towns, villages, and countryside. Exclusion zones were created covering nearly 2,000 square miles (5,000 sq km). These zones have since become havens for wolves, bison, bears, many species of birds, and much more wildlife. However, the extent to which they may have been unharmed by the remains of the radioactive fallout is uncertain.

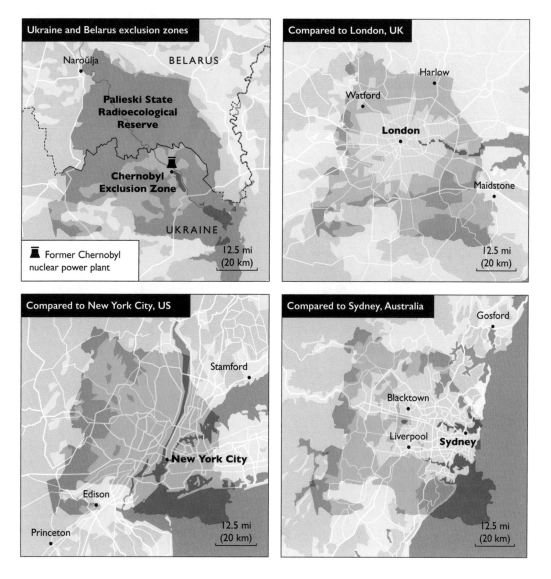

Just a few of the **hundreds of species** that have vanished in the 21st century

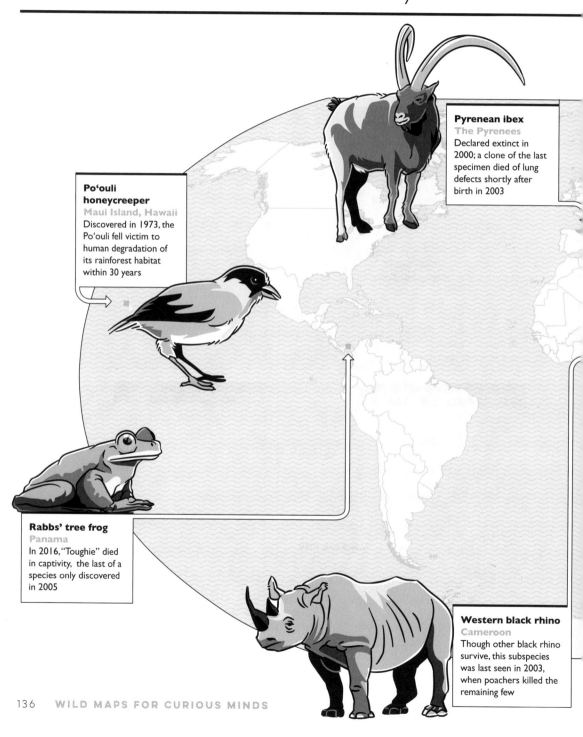

Pyrenean ibex
The Pyrenees
Declared extinct in 2000; a clone of the last specimen died of lung defects shortly after birth in 2003

Po'ouli honeycreeper
Maui Island, Hawaii
Discovered in 1973, the Po'ouli fell victim to human degradation of its rainforest habitat within 30 years

Rabbs' tree frog
Panama
In 2016, "Toughie" died in captivity, the last of a species only discovered in 2005

Western black rhino
Cameroon
Though other black rhino survive, this subspecies was last seen in 2003, when poachers killed the remaining few

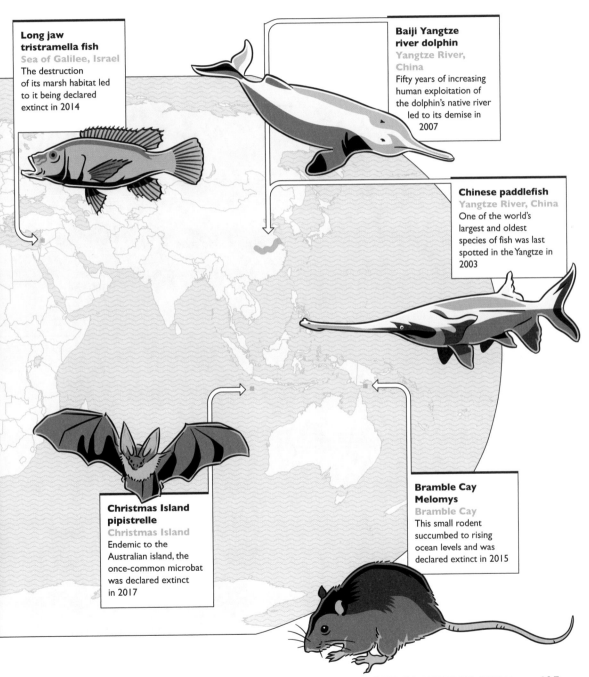

Long jaw tristramella fish
Sea of Galilee, Israel
The destruction of its marsh habitat led to it being declared extinct in 2014

Baiji Yangtze river dolphin
Yangtze River, China
Fifty years of increasing human exploitation of the dolphin's native river led to its demise in 2007

Chinese paddlefish
Yangtze River, China
One of the world's largest and oldest species of fish was last spotted in the Yangtze in 2003

Christmas Island pipistrelle
Christmas Island
Endemic to the Australian island, the once-common microbat was declared extinct in 2017

Bramble Cay Melomys
Bramble Cay
This small rodent succumbed to rising ocean levels and was declared extinct in 2015

The disappearing **Aral Sea**

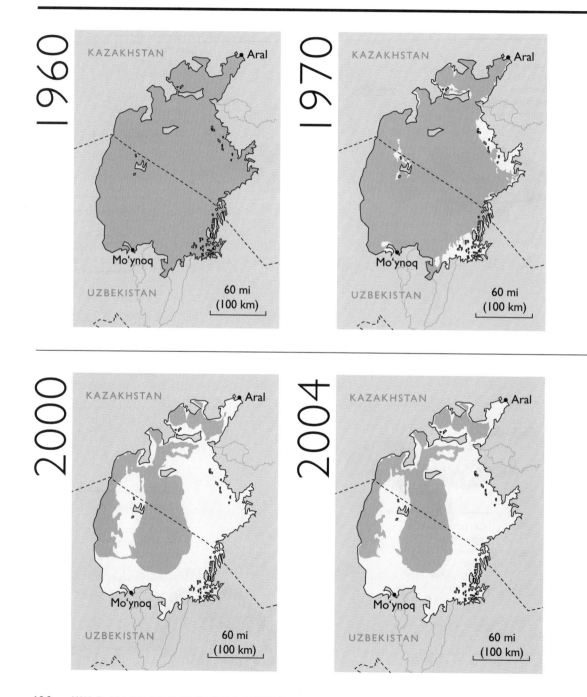

The Aral Sea on the border of Kazakhstan and Uzbekistan was once the world's fourth-largest body of inland water. Its fishing industry was responsible for one sixth of the Soviet Union's entire catch. But by 2002, drainage of the rivers feeding it for the cotton industry caused it shrink in volume by 90 percent. Today it's three times saltier than ordinary seawater and Mo'ynoq, in the past a thriving port, lies more than 45 miles (75 km) from the shoreline.

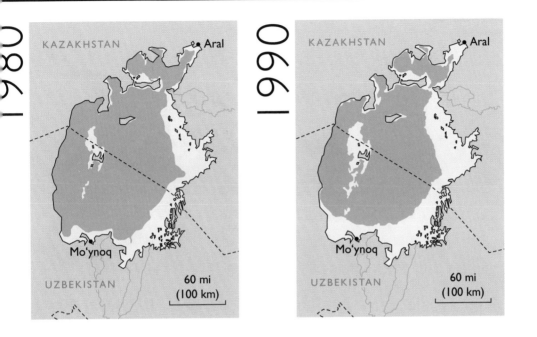

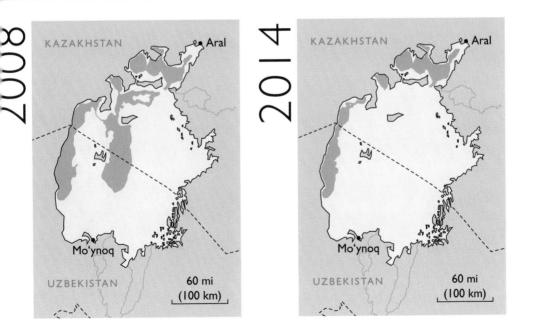

82 Not everyone believes in the **climate crisis**

82

Percentage of those polled who
believe in the climate crisis

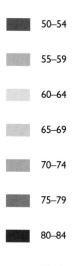

50–54

55–59

60–64

65–69

70–74

75–79

80–84

No data

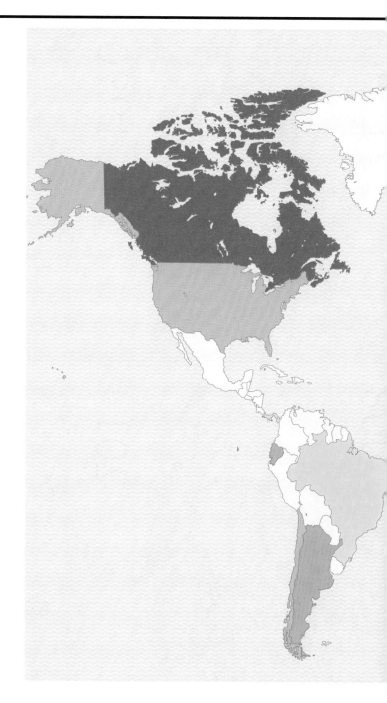

83 How much **forest** have we destroyed?

We have lost one third of the world's forest cover over the last 10,000 years—the equivalent of two thirds of the area of Africa. Half of that loss has taken place over the last century.

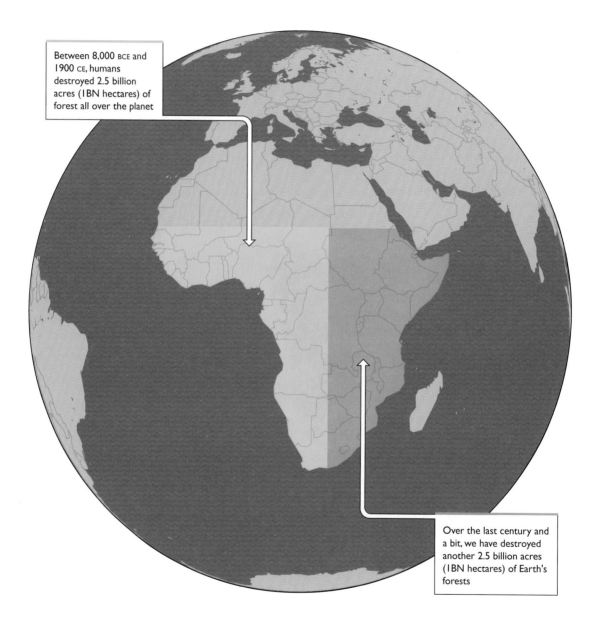

Between 8,000 BCE and 1900 CE, humans destroyed 2.5 billion acres (1BN hectares) of forest all over the planet

Over the last century and a bit, we have destroyed another 2.5 billion acres (1BN hectares) of Earth's forests

84 We are losing **less forest** than before

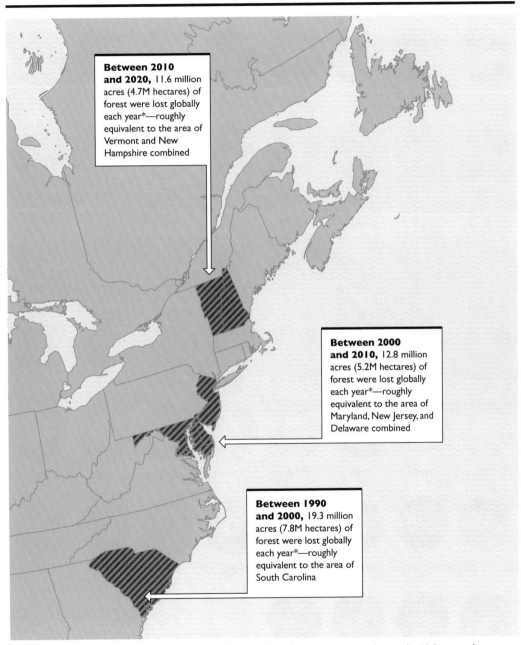

Between 2010 and 2020, 11.6 million acres (4.7M hectares) of forest were lost globally each year*—roughly equivalent to the area of Vermont and New Hampshire combined

Between 2000 and 2010, 12.8 million acres (5.2M hectares) of forest were lost globally each year*—roughly equivalent to the area of Maryland, New Jersey, and Delaware combined

Between 1990 and 2000, 19.3 million acres (7.8M hectares) of forest were lost globally each year*—roughly equivalent to the area of South Carolina

*This figure is calculated by comparing how much forest was lost (to human or natural causes) with how much new forest was created.

A **rapidly warming** world

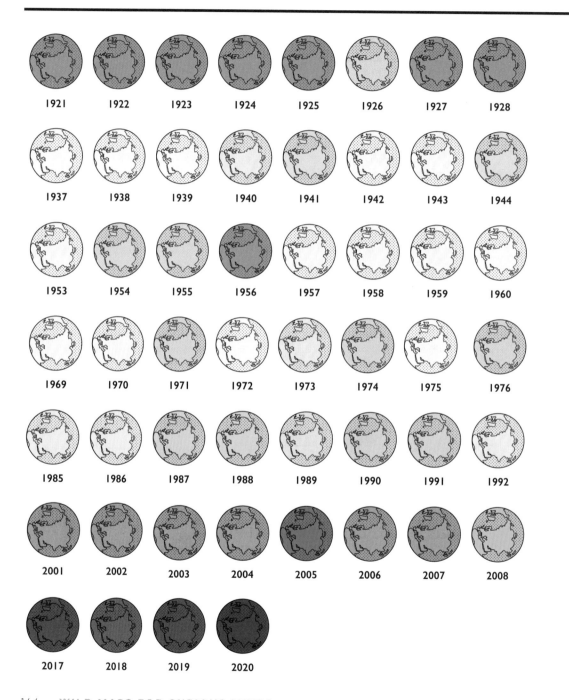

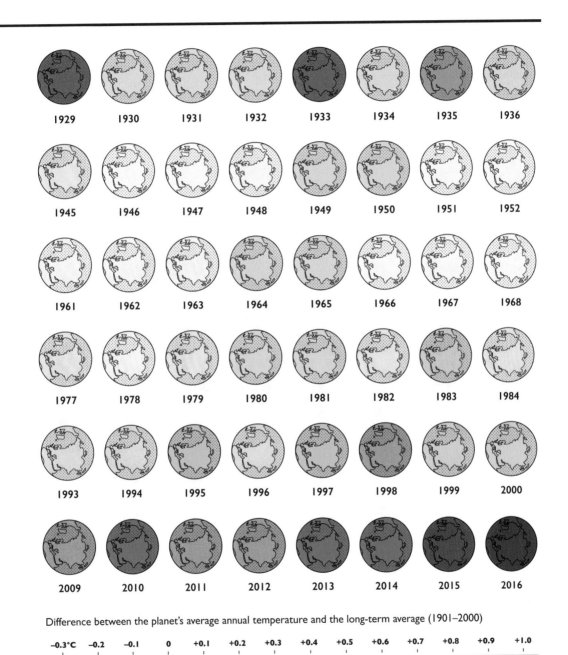

Difference between the planet's average annual temperature and the long-term average (1901–2000)

–0.3°C	–0.2	–0.1	0	+0.1	+0.2	+0.3	+0.4	+0.5	+0.6	+0.7	+0.8	+0.9	+1.0
–0.54°F	–0.36	–0.18	0	+0.18	+0.36	+0.54	+0.72	+0.90	+1.08	+1.26	+1.44	+1.62	+1.8

North America's **summers** are getting longer

Using temperatures in the warmest 90 days of the year, this map compares recent summers (1990–2019) with the baseline period 1960–1989.

Length of summer now (and by how many days it has changed)

80–90 days (10 days or less shorter)   90–95 days (5 days or less longer)   95–100 days (5–10 days longer)

100–105 days (10–15 days longer)   More than 105 days (more than 15 days longer)

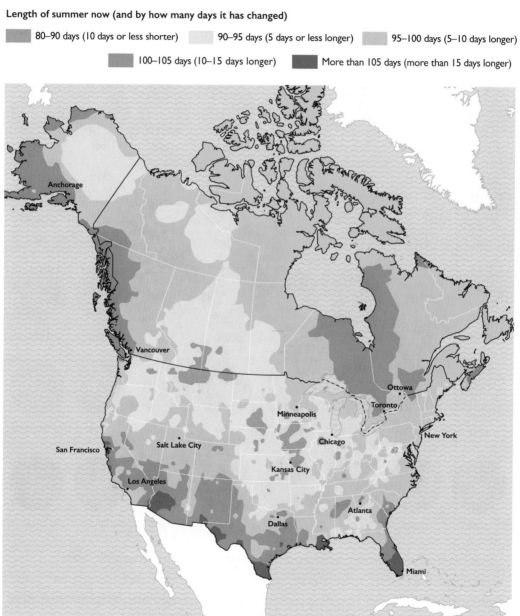

87 …and its **winters** are getting much shorter

Using temperatures in the coldest 90 days of the year, this map compares recent winters (1990–2019) with the baseline period 1960–1989.

Length of winter now (and by how many days it has changed)

- 0–55 days (40–90 days shorter)
- 50–60 days (30–40 days shorter)
- 60–70 days (20–30 days shorter)
- 70–75 days (15–20 days shorter)
- 75–80 days (10–15 days shorter)
- 80–85 days (5–10 days shorter)
- 85–90 days (0–5 days shorter)
- 90–100 days (0–10 days longer)

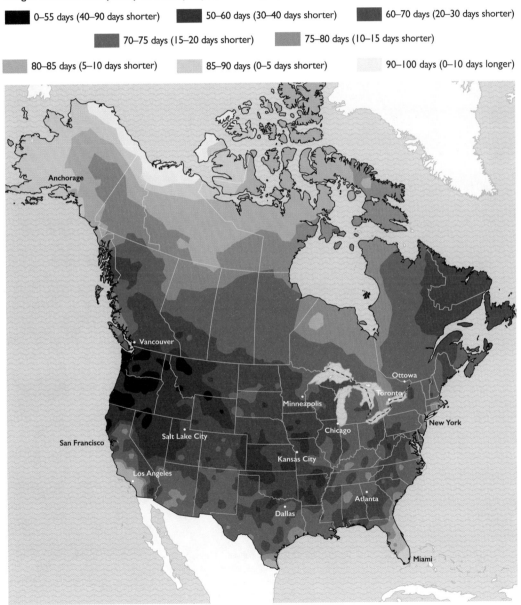

88 Who is at risk from **rising seas?**

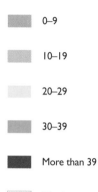

Percentage of population living in areas
16 ft (5 m) above sea level or less

- 0–9
- 10–19
- 20–29
- 30–39
- More than 39
- No data

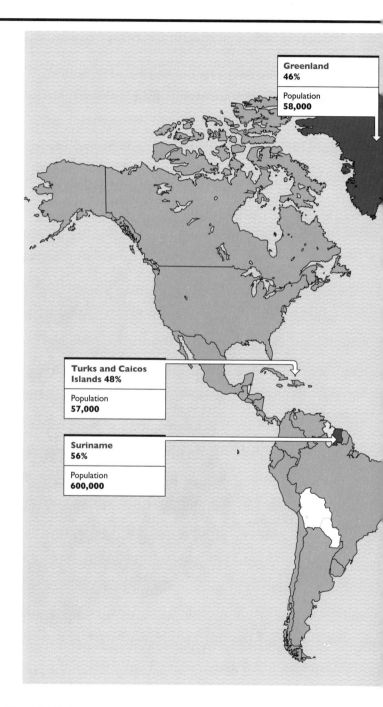

Greenland
46%

Population
58,000

Turks and Caicos
Islands 48%

Population
57,000

Suriname
56%

Population
600,000

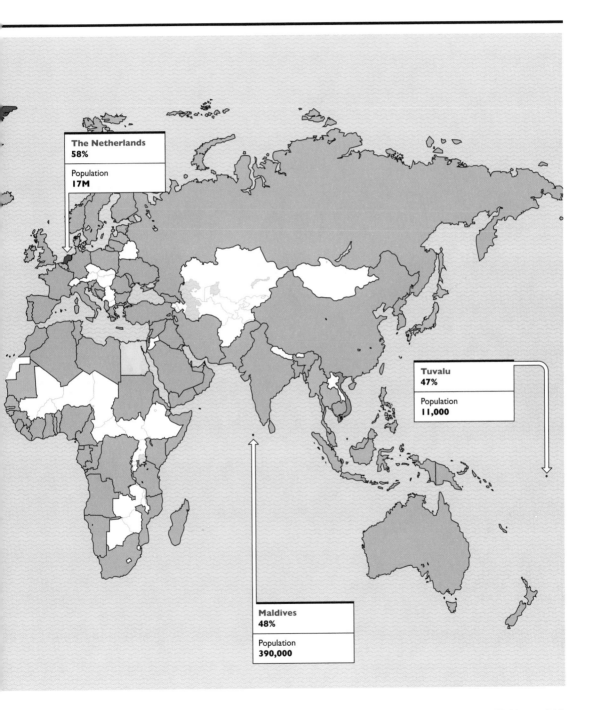

The Netherlands
58%

Population
17M

Tuvalu
47%

Population
11,000

Maldives
48%

Population
390,000

8

THE FINAL FRONTIER

89 All the **forests on Earth** would cover the entire surface of the Moon (and then some)*

Area of Earth's forest cover **17.8 million sq mi** (46.1M sq km)

Area of Earth's surface **197 million sq mi** (510M sq km)

Area of Moon's surface **14.6 million sq mi** (37.9M sq km)

*Distances between stars and planets not to scale

The **70 trees** that have been to the Moon

In 1971, Nasa astronaut Stuart Roosa, a former Forest Service worker, took several hundred tree seeds to the Moon aboard the Apollo 14 mission. Despite being exposed to vacuum at one point, many were germinated back on Earth and planted in the US and beyond.

Type of tree

Sycamore   Loblolly pine   Redwood   Sweet gum   Douglas fir

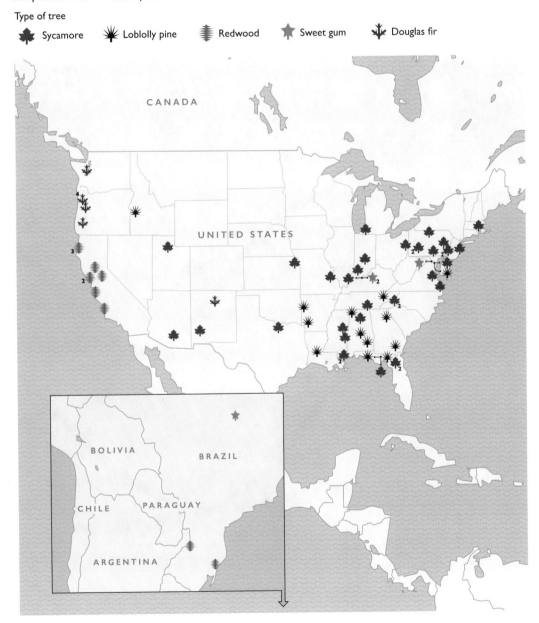

Where to see a **solar eclipse** in the next decade

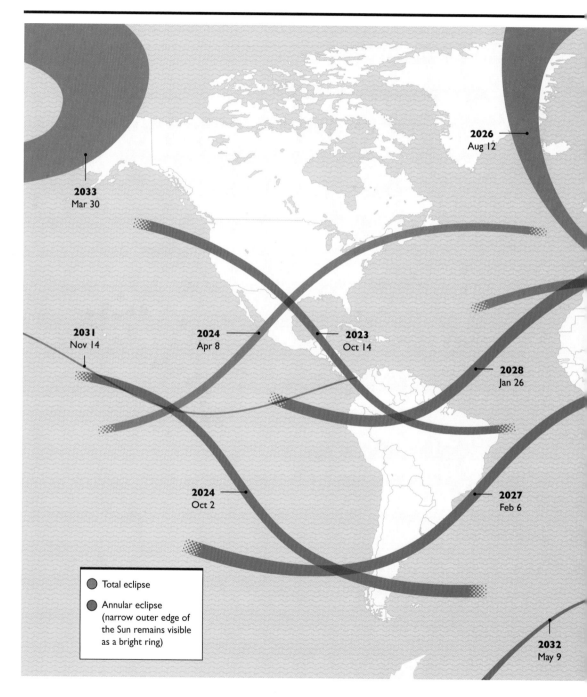

2026
Aug 12

2033
Mar 30

2031
Nov 14

2024
Apr 8

2023
Oct 14

2028
Jan 26

2024
Oct 2

2027
Feb 6

2032
May 9

- Total eclipse
- Annular eclipse (narrow outer edge of the Sun remains visible as a bright ring)

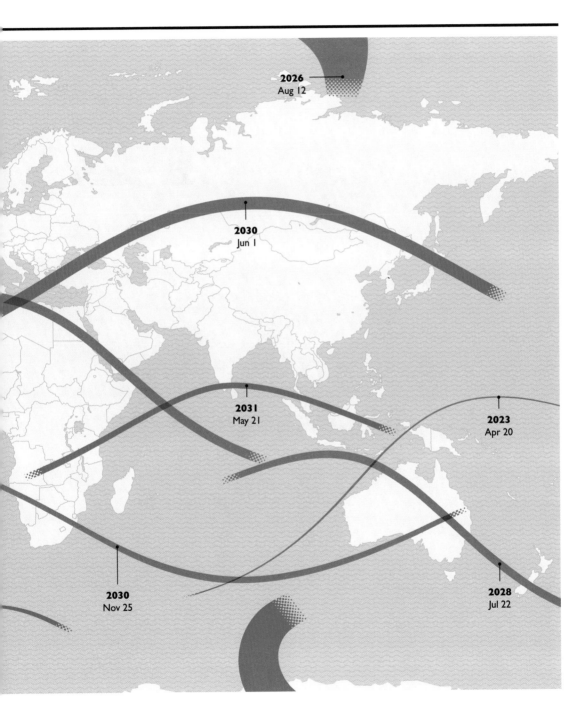

2026
Aug 12

2030
Jun 1

2031
May 21

2023
Apr 20

2030
Nov 25

2028
Jul 22

The Mars volcano **Olympus Mons** is as big as **Arizona**

Over 385 miles (620 km) in diameter and 15 miles (25 km) high, with a caldera 50 miles (80 km) across, this giant Martian volcano is 100 times larger than any on Earth.

Jupiter's **Great Red Spot** could swallow the Earth

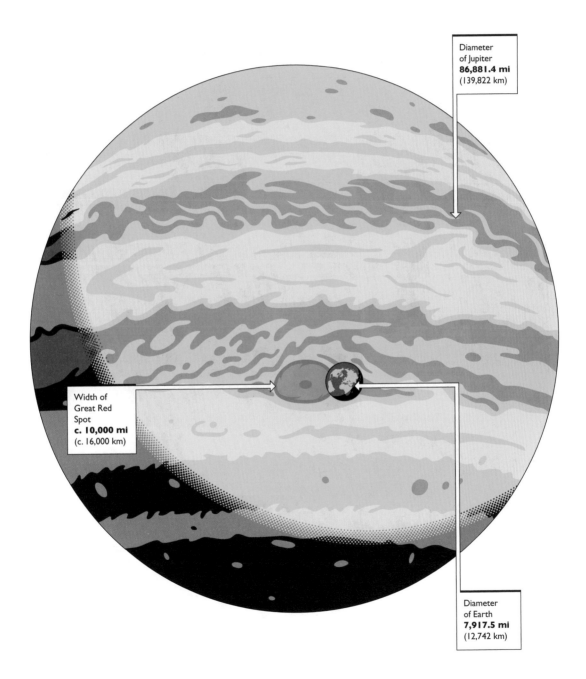

Diameter of Jupiter
86,881.4 mi
(139,822 km)

Width of Great Red Spot
c. 10,000 mi
(c. 16,000 km)

Diameter of Earth
7,917.5 mi
(12,742 km)

94 Guinea pigs **in space!**

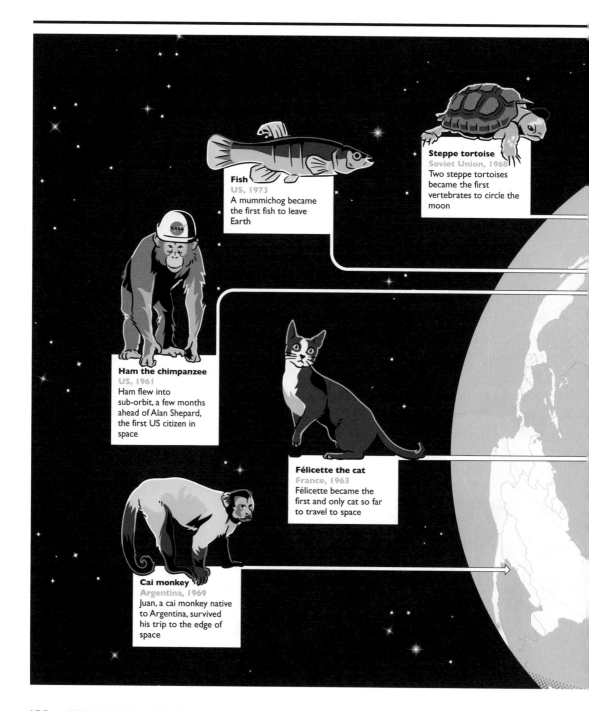

Fish
US, 1973
A mummichog became the first fish to leave Earth

Steppe tortoise
Soviet Union, 1968
Two steppe tortoises became the first vertebrates to circle the moon

Ham the chimpanzee
US, 1961
Ham flew into sub-orbit, a few months ahead of Alan Shepard, the first US citizen in space

Félicette the cat
France, 1963
Félicette became the first and only cat so far to travel to space

Cai monkey
Argentina, 1969
Juan, a cai monkey native to Argentina, survived his trip to the edge of space

Several countries have launched animals into space—here are just a few of the nonhuman astronauts who have boldly gone.

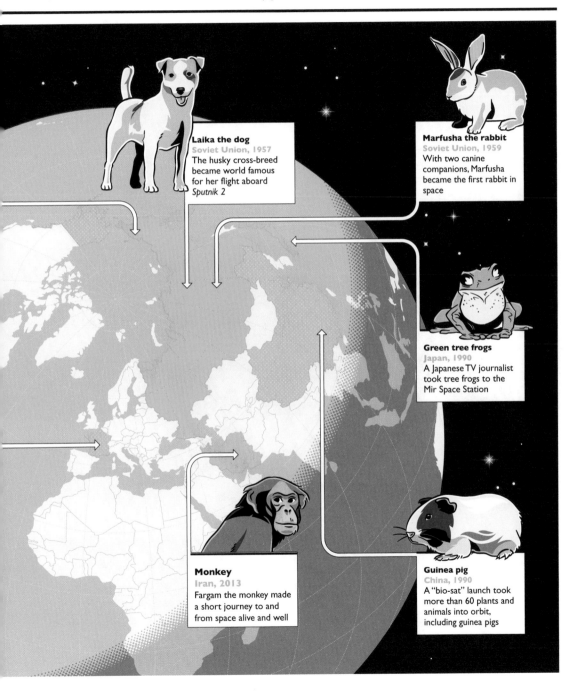

Laika the dog
Soviet Union, 1957
The husky cross-breed became world famous for her flight aboard *Sputnik 2*

Marfusha the rabbit
Soviet Union, 1959
With two canine companions, Marfusha became the first rabbit in space

Green tree frogs
Japan, 1990
A Japanese TV journalist took tree frogs to the Mir Space Station

Monkey
Iran, 2013
Fargam the monkey made a short journey to and from space alive and well

Guinea pig
China, 1990
A "bio-sat" launch took more than 60 plants and animals into orbit, including guinea pigs

95 How big is **Pluto?**

At 1,473 miles (2,370 km) in diameter, the smallest major body in our solar system would fit between London and Moscow.

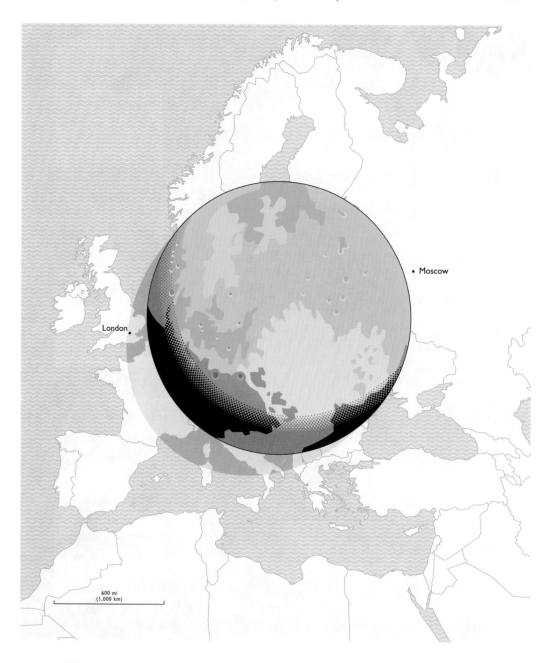

London

Moscow

600 mi
(1,000 km)

96 **Australia** boldly goes

How would the landmass look elsewhere in the solar system?

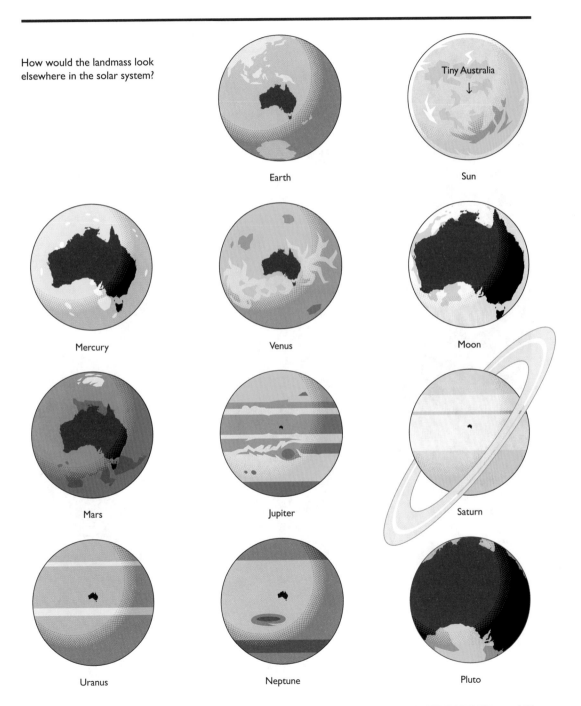

Earth

Sun
Tiny Australia
↓

Mercury

Venus

Moon

Mars

Jupiter

Saturn

Uranus

Neptune

Pluto

97 **All the planets** in our solar system would **fit between the Earth and Moon**

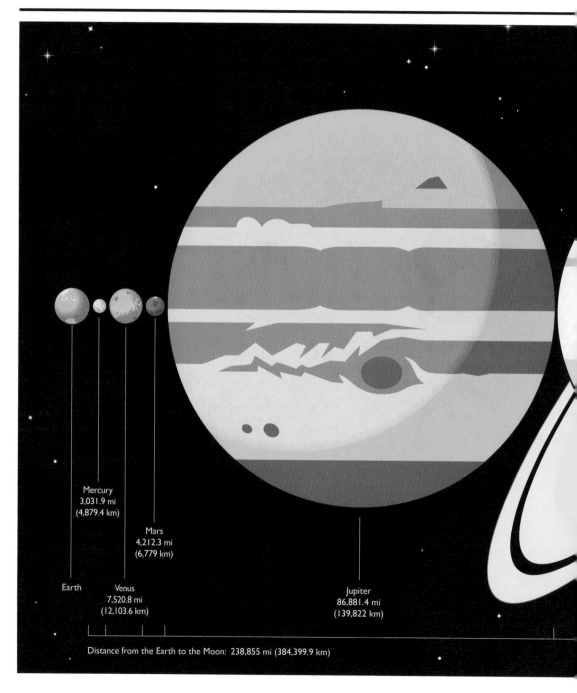

Mercury
3,031.9 mi
(4,879.4 km)

Mars
4,212.3 mi
(6,779 km)

Earth

Venus
7,520.8 mi
(12,103.6 km)

Jupiter
86,881.4 mi
(139,822 km)

Distance from the Earth to the Moon: 238,855 mi (384,399.9 km)

Depending on how you look at it, this astonishing fact either makes the planets seem rather small or makes space seem very, very big. If you lined up every planet in our solar system side by side, the total combined distance, 236,131 miles (375,136.6 km), is almost exactly (but just shy of) the average distance from the Earth to the Moon: 238,855 miles (384,399.9 km). While not officially a planet anymore, there's just enough room left over to squeeze in Pluto, which is 1,473 miles (2,370 km) wide.

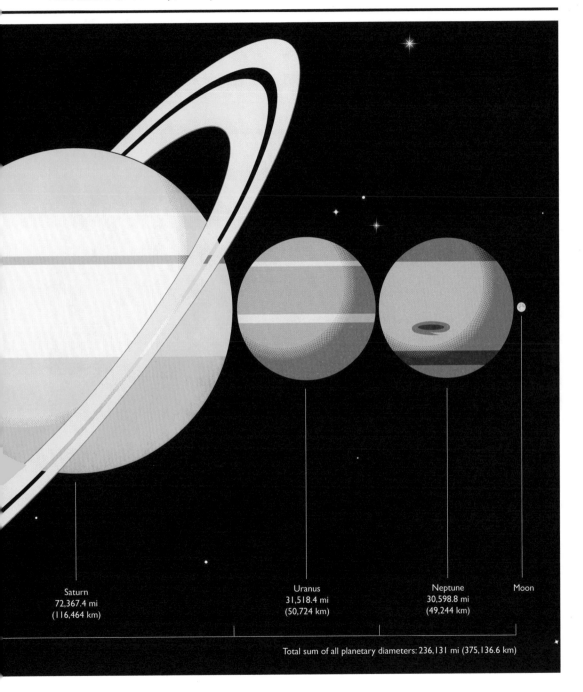

Saturn
72,367.4 mi
(116,464 km)

Uranus
31,518.4 mi
(50,724 km)

Neptune
30,598.8 mi
(49,244 km)

Moon

Total sum of all planetary diameters: 236,131 mi (375,136.6 km)

The rise of the **Luna girls**

In 2003, the name Luna—meaning "moon" in Latin—was America's 889th most popular girls' name. In 2020, it ranked 14th.

Where "Luna" ranked among 2020's most popular baby names

| 1–10 | 11–20 | 21–30 | 31–40 | 41–50 | 51–60 | 61–70 | No data |

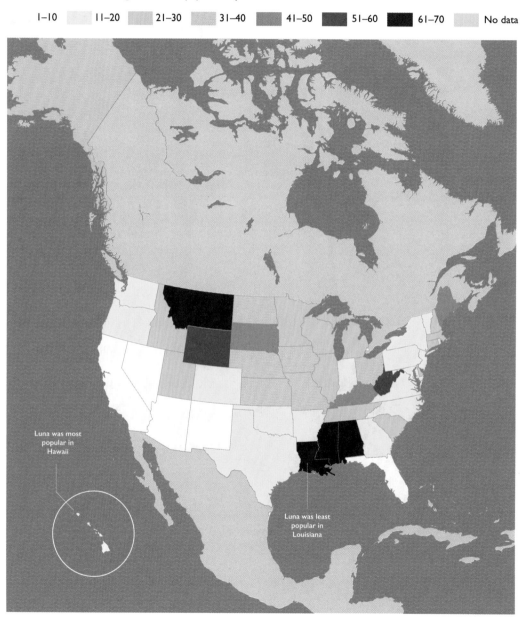

Luna was most popular in Hawaii

Luna was least popular in Louisiana

99 Welcome to **Planet Rock**

Here's how the continents of Earth might look on a planet made up of all the terrestrial planets (Earth, Mars, Venus, Mercury), moons, and other bits of rock floating about our solar system.

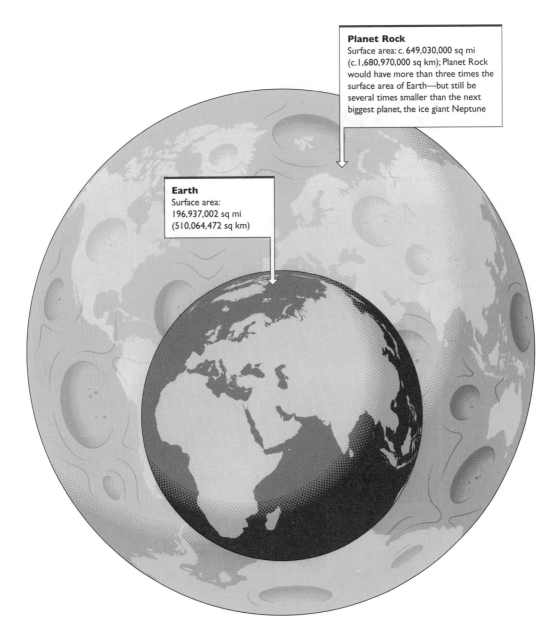

Planet Rock
Surface area: c. 649,030,000 sq mi (c.1,680,970,000 sq km); Planet Rock would have more than three times the surface area of Earth—but still be several times smaller than the next biggest planet, the ice giant Neptune

Earth
Surface area:
196,937,002 sq mi
(510,064,472 sq km)

The **brightest star** in our night sky is a lot bigger than the Sun*

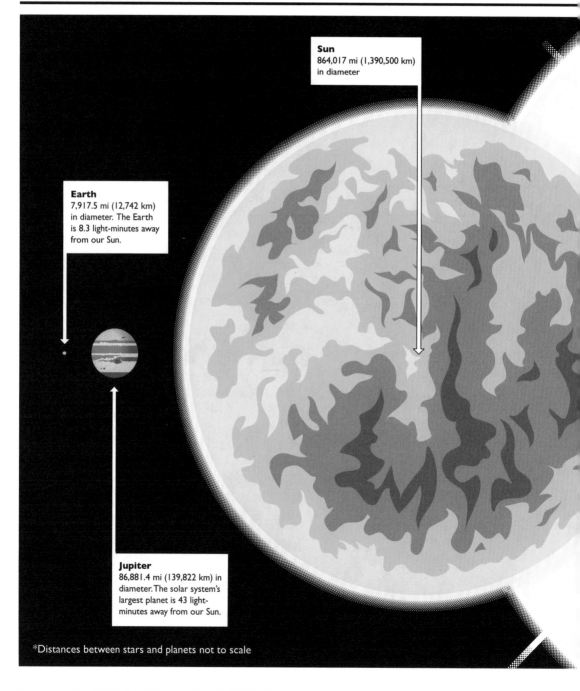

Sun
864,017 mi (1,390,500 km) in diameter

Earth
7,917.5 mi (12,742 km) in diameter. The Earth is 8.3 light-minutes away from our Sun.

Jupiter
86,881.4 mi (139,822 km) in diameter. The solar system's largest planet is 43 light-minutes away from our Sun.

*Distances between stars and planets not to scale

Sirius
1,479,485 mi (2,381,000 km) in diameter. It's 8.6 light-years away from our solar system.

SOURCES

Ancient History

1. Empire of the woolly mammoth
Concept: Adapted from "Woolly Mammoth Late Pleistocene Dymaxion Biogeographic Distribution," Wikimedia Commons author Andrew Z. Colvin, Wikimedia Commons, 2016, under CC BY-SA 4.0.

2. If every human who ever lived crowded into one place
Data: Kaneda, T., Haub, C., "How Many People Have Ever Lived on Earth?" Population Reference Bureau, prb.org; United Nations, Department of Economic and Social Affairs, Population Division (2019): World Population Prospects 2019 © United Nations, under CC BY 3.0 IGO.

3. The spread of humans and the extinction of large mammals
Concept: Adapted from "Human migration and the extinction of large mammals," Hannah Ritchie and Max Roser, OurWorldinData.org, under CC-BY. Data: Andermann et al., "The past and future human impact on mammalian diversity," *Science Advances*, Sept 2020, American Association for the Advancement of Science.

4. North America's supervolcanoes
Concept: "Modeling the Ash Distribution of a Yellowstone Supereruption" (2014), courtesy of the US Geological Survey (USGS), usgs.gov.

5. When the English Channel was a mighty river
Concept and data: Adapted from Figure 13, Patton et al., "Deglaciation of the Eurasian ice sheet complex," *Quaternary Science Reviews*, Aug 1, 2017.

6. The rise and fall of Europe's forests
Concept and data: Modified from Figure 12, Zanon et al., "European Forest Cover During the Past 12,000 Years: A Palynological Reconstruction Based on Modern Analogs and Remote Sensing," *Frontiers in Plant Science* 9, 2018, under CC BY 4.0.

7. Where bears once ruled in Italy
Concept: Adapted from Figures 1 and 2, Tattoni, C. "*Nomen omen*. Toponyms predict recolonization and extinction patterns for large carnivores." *Nature Conservation* 37: 1–16, 2019, under CC-BY.

8. Where wolves once ruled in Italy
Concept: Adapted from Figures 1 and 2, Tattoni, C. "*Nomen omen*. Toponyms predict recolonization and extinction patterns for large carnivores." *Nature Conservation* 37: 1–16, 2019, under CC-BY.

9. The retreat of the hunter-gatherer
Concept: "Extent of foraging, hunting and gathering, 8000 BCE to 1850," Our World in Data, under CC-BY. Data: Stephens et al., "Archaeological assessment reveals Earth's early transformation through land use," *Science*, Aug 30, 2019, American Association for the Advancement of Science.

10. The advance of the farmer
Concept: "Extent of intensive agriculture, 8000 BCE to 1850," Our World in Data, under CC-BY. Data: Stephens et al., "Archaeological assessment reveals Earth's early transformation through land use," *Science*, Aug 30, 2019, American Association for the Advancement of Science.

Out and About

11. Follow your nose: The longest walk in a straight line
Concept: Guy Bruneau. Data: Google Maps.

12. Distribution of the great apes
Concept: a) Adapted from the Great Apes Survival Partnership (GRASP) Great Apes Distribution Map. GRASP was launched in 2001 to ensure the long-term survival of gorillas, chimpanzees, bonobos, and orangutans and their habitat in Africa and Asia. GRASP is a unique alliance of national governments, research institutions, UN agencies, conservation organizations, and private-sector actors. www.un-grasp.org. b) Adapted from "Distribution of the Great Apes (Including Us)," decolonialatlas.wordpress.com. The Decolonial Atlas is an online counter- mapping collective centring Indigenous perspectives, founded in 2014.

13. Not every giraffe looks alike
Concept: Modified from "Genetic subdivision in the giraffe based on mitochondrial DNA sequences," Wikimedia Commons user OldakQuill, Wikimedia Commons 2007, licensed under CC BY 2.0.

14. Where Americans get their wilderness: The top 10 most visited US national parks
Concept: Adapted from "National Parks of the United States by visitor count," a map created by John Nelson of Esri, the mapping software, location intelligence, and spatial analytics company. www.esri.com.
Data: National Park Service, nps.gov.

15. Where cattle, sheep, or pigs outnumber people
Data: FAOSTAT. Crops and livestock products, 2020, FAO.

16. Where to see a giant panda outside China
Data: giantpandaglobal.com

17. Which Americans go on wildlife-watching adventures?
Concept and data: Adapted from "Away-From-Home Wildlife-Watchers by Geographic Region," 2016 National Survey of Fishing, Hunting, and Wildlife- Associated Recreation, US Department of the Interior, US Fish and Wildlife Service, and US Department of Commerce, US Census Bureau.

18. All the private gardens in the UK
Data: Brownbill, A., Dutton, A., "UK natural capital: urban accounts," Aug 8, 2019, under the Open Government Licence v3.0.

19. Where to live if you hate snakes
Concept: "World distribution of snakes," Wikimedia

Commons author Eightofnine, Wikimedia Commons, 2009, under CC0 1.0 Universal Public Domain Dedication.

20. Spoiler alert! Captain Ahab's fatal pursuit of Moby Dick, the Great White Whale
Data: Melville, H., *Moby Dick* (1851).

21. The secret rivers of London
Concept: Adapted from the map "Tributaries of the Thames from Kingston to Erith," Barton, Nicholas, *The Lost Rivers of London* (1st, 2nd, and 3rd editions). This map used by kind permission of Historical Publications, all rights reserved.

22. How fast are you spinning?
Concept: Adapted from "Faster than sound? Our velocity due to Earth's rotation," sciengsustainability. blogspot.com.

The Watery World

23. Imagine all the oceans as a single body of water
Concept: Adapted from "Spilhaus: cool and warm currents," a map created by John Nelson of Esri, the mapping software, location intelligence and spatial analytics company. www.esri.com.
Data: "Major wind driven ocean currents of the world" (2016), www.maps.com.

24. The tiny creek that connects the Atlantic and Pacific
Data: Copyright © Wikipedia, "Parting of the Waters," under CC BY-SA 3.0; River-runner. samlearner.com; Sam Learner is a web and visualization developer from Chicago. He works on gathering, visualizing, and communicating public data in order to make it more accessible. www. samlearner.com Hydro Network-Linked Data Index, USGS; NHDPlus High Resolution, USGS.

25. Where the oceans get their water
Concept: "Ocean drainage," Wikimedia Commons author Citynoise (radicalcartography.net), Wikimedia Commons 2007, under CC0 1.0.

26. Point Nemo: The most remote place on Earth
Concept: Adapted from "Oceanic pole of inaccessibility," Wikimedia Commons author Timwi, Wikimedia Commons 2007, under CC0 1.0.

27. Lighthouses of Britain
Concept: Adapted from "GB lighthouses and range of their light," Steer. Design for Movement at Steer are a team of designers, cartographers, geospatial analysts, architects, and wayfinding planners who deliver work across the world to enhance our cities, transport, and public places. dfm.steergroup.com. Data: Copyright © Wikipedia "List of lighthouses in England," "List of lighthouses in Wales," "List of lighthouses in Scotland," under CC BY-SA 3.0.

28. The soggiest places in North America
Data: "NOAA National Centers for Environmental Information: State Climate Summaries 2022," 2022, NOAA Technical Report NESDIS 150; "Weather Extremes in Canada," Wikimedia Commons, under CC BY-SA 3.0; "Mexico," Climate Change Knowledge Portal from World Bank Group; Giambelluca, T. W., et al., "Online Rainfall Atlas of Hawai'i," Bull. Amer. Meteor. Soc. 94 (2013): 313–16.

29. Which is the wettest half of the year in North America?
Concept: "Which is the Wettest Half of the Year: Apr–Sept, Oct–Mar," Brian Brettschneider, us-climate.blogspot.com.
Data: US Climate Normals, NOAA; Global Historical Climatology Network monthly V4, NOAA.

30. It's raining, but is it pouring?
Concept: Erin Davis, "Does it rain? Does it pour?" Davis is a data visualization developer living in Portland, Oregon, US. She specialises in making maps to reveal the hidden patterns in the world around us. www.erdavis.com.
Data: Hersbach, H., et al., "ERA5 hourly data on single levels from 1979 to present" (2018). Copernicus Climate Change Service (C3S) Climate Data Store.

31. Level of water stress across North America
Source: WRI Aqueduct, accessed on June 30, 2022, aqueduct.wri.org, under CC BY-SA 4.0.

32. Where it might snow on Christmas Day across North America
Concept: "Historical Probability of Measurable Snow on Christmas Day," Brian Brettschneider, us-climate. blogspot.com.
Data: Global Historical Climatology Network monthly, NOAA.

Geography

33. These spheres represent all the water and air on Earth
Concept and data: Adam Nieman, a Bristol-based artist who mainly works with science. He has a degree in physics and a PhD in the visual culture of science. He is the founder and creative director of Real World Visuals, a data visualization company that specialises in "concrete visualisation"—turning numbers into actual stuff. With thanks to Howard Perlman of USGS.

34. How deep is the Earth?
Original concept.

35. The middles of nowhere
Data: Copyright © Wikipedia, "Pole of inaccessibility," under CC BY-SA 3.0.

36. Which countries lie directly east and west of the Americas?
Concept: Eric Odenheimer, who is a lover of college football, film, maps, statistics, and any and all peculiar trivia therein. You can find more of his work on Reddit under the username e8odie.

37. What Antarctica looks like beneath the ice
Concept: Modification of "Antarctic Bedrock," Wikimedia Commons author Paul V. Heinrich, Wikimedia Commons, 2008, under CC BY-SA 3.0.

38. The nations with no sea view
Concept: Adapted from "Landlocked countries: 42 landlocked (green), 2 doubly landlocked (purple)," Wikimedia Commons author NuclearVacuum, Wikimedia Commons 2011, under CC BY-SA 3.0.

39. All the rivers in the world
Concept: "HydroRIVERS: Global river network at high spatial resolution,"
© HydroLAB. Bernhard Lehner is a professor of global hydrology at McGill University in Montreal, Canada. Putting water on a map is both his passion and his profession. Over the last 20 years, he and his research team have created digital global maps of everything that flows, drains, drips, soaks, or pools, from the world's rivers to lakes, watersheds, wetlands, reservoirs, and even waterfalls.
Data: Lehner, B., Verdin, K., Jarvis, A., "New global hydrography derived from spaceborne elevation data. Eos, Transactions," AGU 89, no. 10: 93–94, 2008.

40. The surprisingly few countries named after animals

Data: Copyright © Wikipedia, "List of country-name etymologies," under CC BY-SA 3.0.

41. The hemisphere of water

Concept: Adapted from "Hemisphere water," Wikimedia Commons author Citynoise (radicalcartography.net), Wikimedia Commons 2015, under CC BY-SA 4.0.

42. The hemisphere of land

Concept: Adapted from "Hemisphere land," Wikimedia Commons author Citynoise (radicalcartography.net), Wikimedia Commons 2015, under CC BY-SA 4.0.

43. All the lakes in the world

Concept: "HydroLAKES: 1.4 million global lakes (10ha or larger)," © HydroLAB.
Biography and data: same as for "All the rivers in the world."

44. How the natural world flies its flag

Data: Copyright © Wikipedia, "List of national flags by design," under CC BY-SA 3.0.

45. Here be dragons

Original concept.

46. The state of land use

Data: "GAO-20-461R Mining on Federal Lands," U.S. Government Accountability Office, 2020; "Golf Course Environmental Profile," Golf Course Superintendents Association of America, Copyright © 2007; Frazer, L., "Paving Paradise: The Peril of Impervious Surfaces," *Environmental Health Perspectives*, Vol 113, 2005; Samuel Stebbins, "Who Owns the Most Land in America? Jeff Bezos and John Malone Are Among Them," *USA Today*, 2019; Krystal D'Costa, "The American Obsession with Lawns," *Scientific American*, 2017; Center for Sustainable Systems, University of Michigan, 2021, "U.S. Cities Factsheet," Pub. No. CSS09-06; "Land Cover Change," NOAA Office for Coastal Management, Coastal Change Analysis Program, Regional Land Cover 1996 to 2016; Lower, R. and Watson, R., "How Many National Parks Are There?" National Park Foundation, 2020; "State Parks," Encyclopedia of Recreation and Leisure in America (Encyclopedia.com), 2022; Water Science School, "How Wet Is Your State? The Water Area of Each State," USGS, 2018; Bigelow, D. P. and Borchers, A., "Major Issues of Land in the United States, 2012," United States Department of Agriculture, 2017.

Using and Abusing Nature

47. How much arable land is there?

Data: The CIA Factbook, cia.gov; World Bank, data.worldbank.org.

48. The windy stretches that could help power the world

Concept: "Ocean Area Required to Power the World with Zero Carbon Emissions using only Offshore Wind," 2020, Land Art Generator initiative (LAGi). LAGi consults with city planners, developers, and communities to install renewable energy installations that beautify public places as works of civic art and inspire people about the benefits of decarbonization. Find more maps and information graphics at landartgenerator.org.
Data: Global Wind Atlas 3.0, under CC BY 4.0; "Capacity Densities of European Offshore Wind Farms," Deutsche WindGuard GmbH, 2018; "Global energy transformation: A roadmap to 2050" (2019 edition), © International Renewable Energy Agency (IRENA); with thanks to the Federal Maritime and Hydrographic Agency (www.bsh.de).

49. The hotspots in Earth's crust that could help power the world

Concept: Adapted from Figure 2, J. M. K. C. Donev et al., "Geothermal Electricity—Energy Education," 2021, energyeducation.ca, under CC BY SA. Data: Wolfson, R., "Energy from Earth and Moon" in *Energy, Environment, and Climate*, 2nd ed., New York, NY: W.W. Norton & Company, 2012, ch. 8, pp. 204–224; World Bank, data.worldbank.org
Data: Tokyo Institute of Technology, "Could the heat of the Earth's crust become the ultimate energy source?" *ScienceDaily*, 2019.

50. The sunny places that could help power the world

Concept: Adapted from "Photovoltaic Power Potential," © 2020, The World Bank.
Data: Global Solar Atlas 2.0, under CC BY 4.0; Solargis; "Global energy transformation: A roadmap to 2050" (2019 edition), International Renewable Energy Agency, © IRENA.

51. National treasures: The biggest producers of gem diamonds in carats

Data: "World Gem Diamond Mine Production (2019)," USGS, Mineral Commodity Summaries, 2021.

52. Who are the gold diggers?

Data: "Gold mine production (2019)," Gold Production, Gold Hub. Used with permission from the World Gold Council.

53. China grows 25 million tons of garlic annually
Data: FAOSTAT, Crops and livestock products, 2019, Food and Agriculture Organization of the United Nations (FAO).

54. Who is behind the great avocado boom?
Data: FAOSTAT, Crops and livestock products, 2000 & 2019, FAO.

55. Who eats their greens?
Concept: "Vegetable supply per capita, 2017," Our World in Data, under CC BY.
Data: FAOSTAT, Food Balances, 2019, FAO.

56. Who eats the most fruit?
Concept: "Fruit consumption per capita, 2017," Our World in Data, under CC BY.
Data: FAOSTAT, Food Balances, 2019, FAO.

57. Who eats the most meat?
Concept: "Meat supply per person, 2017," Our World in Data, under CC BY 4.0.
Data: Source: FAOSTAT, Food Balances, 2019, FAO.

58. The dairy lovers
Data: FAOSTAT, Food Balances, 2019, FAO.

59. Where do all the turkeys live?
Data: FAOSTAT, Crops and livestock products, 2020, FAO.

60. Who has killed whales since the 1985 ban?
Data: International Whaling Commission.

61. Where do America's hunters live?
Concept and data: Adapted from 2016 National Survey of Fishing, Hunting, and Wildlife-Associated Recreation, US Department of the Interior, US Fish and Wildlife Service, and US Department of Commerce, US Census Bureau.

62. Where rhinos have been poached
Concept: Adapted from "Number of rhinos poached, 1990 to 2017," Our World in Data, under CC-BY.
Data: Emslie, R.H., et al., "African and Asian rhinoceroses—status, conservation and trade," 2019, a report from the IUCN Species Survival Commission African and Asian Rhino Specialist Groups and TRAFFIC to the CITES Secretariat pursuant to Resolution Conf. 9.14 (Rev. CoP17). Report to CITES 17th meeting (Colombo, June 2019); www.poachingfacts.com.

63. Where American ships killed whales in the 19th century
Concept: Peter Atwood, as above.
Data: American Offshore Whaling Voyages: A Database, New Bedford Whaling Museum and Mystic Seaport Museum, Inc. (whalinghistory.org/av), under CC BY 4.0.

64. Who are the world's cat people?
Data: "Which countries have the most cat owners? Share of people who own a cat in selected countries in 2017," Dalia Research, via Statista.com.

65. Which countries are responsible for the most types and breeds of dog?
Data: Copyright © Wikipedia, "Dog breeds by country of origin," under CC BY 3.0.

66. Who has kept dolphins in captivity?
Concept and data: "Behind the smile: The multi-billion-dollar dolphin entertainment industry," World Animal Protection, 2019.

Extreme Earth

67. The oldest living trees in the world
Data: Copyright © Wikipedia, "List of oldest trees," under CC BY 3.0.

68. The highs and lows of North America
Data: Nag, Oishimaya Sen, "The Tallest Peaks In North America," WorldAtlas, 2017; "Badwater Basin," National Park Service (www.nps.gov), 2021; "Salton Trough," Wikimedia Commons, under CC BY-SA 3.0; "Lake Enriquillo," Wikimedia Commons, under CC BY-SA 3.0; "List of U.S. states and territories by elevation," Wikimedia Commons, under CC BY-SA 3.0; "Alberta," Copyright © by Team The World of Info; "Canada," Topographic Maps, source data from: Yamazaki, D., D. Ikeshima, R. Tawatari, T. Yamaguchi, F. O'Loughlin, J. C. Neal, C. C. Sampson, S. Kanae, and P. D. Bates, "A High Accuracy Map of Global Terrain Elevations," *Geophysical Research Letters* 44 (2017).

69. Warning! Dangerous animals at work
Concept: "Death rate from venomous animal contact, 2019," Our World in Data, under CC BY 4.0.
Data: Global Burden of Disease Collaborative Network, "Global Burden of Disease Study 2019 (GBD 2019) Results," Seattle, United States, Institute for Health Metrics and Evaluation (IHME), 2021. All rights reserved.

70. Watch your step—these creatures are the tiniest of their kind
Data: National Geographic; Wikipedia; Nature; BBC; fieldguide.mt.gov.

71. How far are you from an earthquake zone?
Concept and data: "The Global Earthquake Model (GEM) Global Seismic Hazard Map (version 2018.1)." The Global Earthquake Model Foundation is a non-profit, public-private partnership that drives a global collaborative effort to develop scientific and high-quality resources for transparent assessment of earthquake risk and to facilitate their application for risk management around the globe. www.globalquakemodel.org/gem.

72. Where the wild places still are
Data: "Last of the Wild, v2," Socioeconomic Data and Applications Center (SEDAC), 2004; EarthData, earthdata.nasa.gov.

73. Where are people at most risk from natural disasters?
Concept: Adapted from "Exposure of the population to the natural hazards earthquakes, cyclones, floods, droughts, and sea-level rise," World Risk Report 2021 © Bündnis Entwicklung Hilft 2021.
Data: Center for Remote Sensing of Ice Sheets; Human Footprint maps, Center for International Earth Science Information Network (CIESIN), under CC BY-SA 4.0; IFHV; Oak Ridge National Laboratory Landscan.

74. Ocean deep, mountain high
Concept: Peter Atwood, a mapmaker and designer from Halifax, Canada, peteratwoodprojects.wordpress.com.

75. The largest iceberg reliably recorded was bigger than Corsica and Mallorca
Data: Antarctic Meteorological Research Center, Space Science and Engineering Center, University of Wisconsin-Madison; Copyright © Wikipedia, "List of recorded icebergs by area," under CC BY 3.0.

76. How people have died in America's national parks
Data: "Danger Parks Ranked," Outforia.com, via Freedom of Information request to the National Park Service (www.nps.gov).

77. All the F5/EF5 tornadoes reliably recorded
Data: "List of F5 and EF5 tornadoes," Wikimedia Commons, under CC BY-SA 3.0; NOAA National Weather Service, Storm Prediction Center, spc.noaa.gov.

78. How many acres burn in the US each year?
Data: "Climate Change Indicators: Wildfires," EPA; NIFC, 2021; USDA Forest Service, 2014.

79. The Chernobyl disaster led to a nature reserve six times the size of New York City
Concept: Adapted from "Chernobyl Exclusion Zone," Wikimedia Commons author Nzeemin, Wikimedia Commons 2012, under CC BY-SA 2.0. Adapted from "Polesie State Radioecological Reserve (OpenStreetMap)," Wikimedia Commons creator OpenStreetMap contributors, under CC BY-SA 2.0.

80. Just a few of the hundreds of species that have vanished in the 21st century
Data: Lifegate.com, *Scientific American*, International Union for Conservation of Nature, BBC, Save the Rhino, *New Scientist*, US Fish and Wildlife Service.

81. The disappearing Aral Sea
Concept: Adapted from "Aral Sea," Wikimedia Commons author NordNordWest, Wikimedia Commons 2008, under CC BY-SA 3.0.

82. Not everyone believes in the climate crisis
Data: Flynn et al., "Peoples' Climate Vote: Results," © UNDP and University of Oxford, 2021.

83. How much forest have we destroyed?
Data: Our World in Data, under CC BY; Williams, M. (2003), *Deforesting the Earth: From Prehistory to Global Crisis*, University of Chicago Press; FAO and UNEP, 2020, "The State of the World's Forests 2020"; FAOSTAT; Ellis, E. C., Beusen, A. H., & Goldewijk, K. K. (2020), "Anthropogenic Biomes: 10000 BCE to 2015 CE," Land 9, no 5, 129, under CC BY 4.0.

84. We are losing less forest than before
Data: "Global Forest Resources Assessment 2020—Key Findings," Rome, copyright © FAO, 2020.

85. A rapidly warming world
Data: NOAA National Centers for Environmental Information, Climate at a Glance: Global Time Series.

86. North America's summers are getting longer
Concept: "How Long is Summer Compared to What It Used to Be?" (US), Brian Brettschneider, us-climate.blogspot.com.
Data: Menne, M. J., Williams, C. N., Gleason,

B. E., Jared Rennie, J., Lawrimore, J. H., "The Global Historical Climatology Network Monthly Temperature Dataset, Version 4."

87. . . . and its winters are getting much shorter
Concept: "How Long is Winter Compared to What It Used to Be?" (US), Brian Brettschneider, us-climate.blogspot.com.
Data: same as for "North America's summers are getting longer."

88. Who is at risk from rising seas?
Concept: Adapted from "Population living in areas where elevation is below 5 meters (% of total population)," The World Bank, under CC-BY 4.0.
Data: "Low Elevation Coastal Zone (LECZ) Urban-Rural Population and Land Area Estimates, Version 2," 2013. NASA Socioeconomic Data and Applications Center (SEDAC), CIESIN, under CC BY 4.0.

The Final Frontier

89. All the forests on Earth would cover the entire surface of the Moon (and then some)
Data: "Global Forest Resources Assessment 2020—Key Findings," © FAO.

90. The 70 trees that have been to the Moon
Data: NSSDCA/NASA, with thanks to Dave Williams.

91. Where to see a solar eclipse in the next decade
Concept: Courtesy of Fred Espenak, NASA/Goddard Space Flight Center, eclipse.gsfc.nasa.gov.

92. The Mars volcano Olympus Mons is as big as Arizona
Data: NASA.

93. Jupiter's Great Red Spot could swallow the Earth
Concept and data: NASA.

94. Guinea pigs in space!
Data: Copyright © Wikipedia, "Animals in space," under CC BY 3.0.

95. How big is Pluto?
Original concept.

96: Australia boldly goes
Concept and data: Engaging Data. engaging-data.com is a website for anyone who is curious about the world. The goal is to educate, amuse, and inspire using

data with interactive maps and visualization tools. Its creator, Chris Yang, is a research scientist focusing on energy and environmental issues, and lives in California.

97. All the planets in our solar system would fit between the Earth and Moon
Concept: Thomas Verez
Data: NASA Science, Solar System Exploration, solarsystem.nasa.gov; NASA Science, Space Place, spaceplace.nasa.gov; NASA, New Horizons, nasa.gov.

98. The rise of the Luna girls
Concept: Phillip Reese, a data specialist at The Sacramento Bee and an assistant professor of journalism at Sacramento State. His journalism has won the George Polk and Worth Bingham awards, and he was a finalist for the 2014 Pulitzer Prize for Investigative Reporting.
Data: "Popular Baby Names, 2020," Social Security Administration, ssa.gov.

99. Welcome to Planet Rock
Data: Copyright © Wikipedia, "Solar System objects by size," under CC BY-SA 3.0.

100. The brightest star in our night sky is a lot bigger than the Sun
Data: Copyright © Wikipedia, "Sirius," under CC BY-SA 3.0.

ABOUT THE AUTHORS

MIKE HIGGINS has more than twenty years' experience editing and writing for news media organizations and publishers such as *The Independent*, *The Guardian*, *The Telegraph*, Lonely Planet, and Condé Nast. He lives in south London.

MANUEL BORTOLETTI is an award-winning freelance graphic designer focused on editorial design, infographics, illustration, and art direction. He lives in Venice, Italy.

manuelbortoletti.com

IAN WRIGHT, consultant on this book, runs Brilliant Maps, one of the most popular cartographic sites on the internet, which inspired the Maps for Curious Minds series. Originally from Canada, he now lives in the UK.

brilliantmaps.com

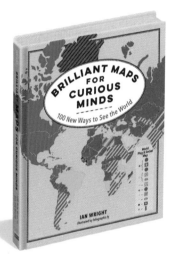
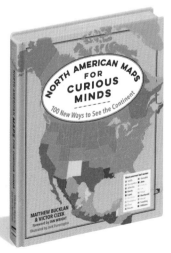